Springer Water

The book series Springer Water comprises a broad portfolio of multi- and interdisciplinary scientific books, aiming at researchers, students, and everyone interested in water-related science. The series includes peer-reviewed monographs, edited volumes, textbooks, and conference proceedings. Its volumes combine all kinds of water-related research areas, such as: the movement, distribution and quality of freshwater; water resources; the quality and pollution of water and its influence on health; the water industry including drinking water, wastewater, and desalination services and technologies; water history; as well as water management and the governmental, political, developmental, and ethical aspects of water.

More information about this series at http://www.springer.com/series/13419

Anja du Plessis

Water as an Inescapable Risk

Current Global Water Availability, Quality and Risks with a Specific Focus on South Africa

Anja du Plessis
Department of Geography, School of
Ecological and Human Sustainability
University of South Africa
Johannesburg, South Africa

ISSN 2364-6934　　　　　　　ISSN 2364-8198　(electronic)
Springer Water
ISBN 978-3-030-03185-5　　　ISBN 978-3-030-03186-2　(eBook)
https://doi.org/10.1007/978-3-030-03186-2

Library of Congress Control Number: 2018960470

© Springer Nature Switzerland AG 2019
This work is subject to copyright. All rights are reserved by the Publisher, whether the whole or part of the material is concerned, specifically the rights of translation, reprinting, reuse of illustrations, recitation, broadcasting, reproduction on microfilms or in any other physical way, and transmission or information storage and retrieval, electronic adaptation, computer software, or by similar or dissimilar methodology now known or hereafter developed.
The use of general descriptive names, registered names, trademarks, service marks, etc. in this publication does not imply, even in the absence of a specific statement, that such names are exempt from the relevant protective laws and regulations and therefore free for general use.
The publisher, the authors and the editors are safe to assume that the advice and information in this book are believed to be true and accurate at the date of publication. Neither the publisher nor the authors or the editors give a warranty, express or implied, with respect to the material contained herein or for any errors or omissions that may have been made. The publisher remains neutral with regard to jurisdictional claims in published maps and institutional affiliations.

This Springer imprint is published by the registered company Springer Nature Switzerland AG
The registered company address is: Gewerbestrasse 11, 6330 Cham, Switzerland

Preface

Water has been and continues to be recognised as a vital resource for all aspects of human and environmental health and life. The wide reality is, even though it has been described as a fundamental human right and international goals have been developed around achieving improved access and sustainability of it, water is still continuously being degraded and misused by all water use sectors around the world.

The combination of a constant increase in the world's human population, continued desire for economic growth as well as continued water pollution will lead to intensified competition between the different water usage sectors and place more pressure on the world's water resources as these sectors attempt to satisfy demands. Additionally, the increasing threat of climate variability, water scarcity and stress around the world, specifically in South Africa, will also put added pressure on water availability and consequently create increased competition between water sectors as each demand more with the continued increase in population and economic growth. A "Business as Usual" mentality is, therefore, unsustainable and new "business unusual" strategies need to be developed.

Water is recognised as an inescapable risk for all spheres due to unsustainable water use, continued water degradation and the evident threat of climate variability around the world and more specifically in South Africa. The book is consequently divided into two parts, namely "Water as a Global Risk" and "Water as a Regional and National Risk" in the context of Southern and South Africa to exemplify the various risks that water problems currently pose now and in the future to all spheres, i.e. the environment, society and economy.

Focus is placed on current and future water availability and scarcity with the use of case studies to obtain a better understanding of the state of the world's water resources and emphasise the increasing challenges that water scarcity and stress pose around the globe. The influences of climate variability on water availability is also recognised through illustrating current observations, impacts, vulnerabilities and future risks. Case studies are used to highlight that the effects of climate variability are already a reality in various countries around the world and affect the environment, human settlements, health and livelihoods as well as the economy.

This book further emphasises that the degradation of water quality should be recognised as a major threat and risk around the world. Poor water quality directly translates into decreasing the availability of the world's water resources and attributes to water scarcity. Water scarcity caused by increased poor water quality has been exacerbated by increased pollution mainly caused by the disposal of large quantities of insufficiently treated, or untreated, wastewater. Chapter 5 of the book consequently illustrates this by focussing on the primary water quality challenges, main contaminants and the dirtiest places around the world to ultimately highlight the dire situation which some regions currently face.

Part II of the book focusses on evaluating South Africa's water resources as a whole in terms of its fresh surface water resources' quality. A brief background of Southern Africa's water resources (surface and groundwater) is provided followed by the establishment of current water quality risk areas for the nine Water Management Areas of South Africa. The establishment of water quality as well as its overall risk for the whole of South Africa is of prime importance as it attempts to create and contribute to informed knowledge and awareness for current and future water uses. These established results should be used to improve current water management practices and decision-making processes.

South Africa's National Water Resource Strategy of 2014 acknowledged that the resource is not receiving the attention and status it deserves and wastage, pollution and degradation is widespread. The sustainability of South Africa's freshwater resources has reached a critical point. Real opportunities exist where South Africa can emerge as a leader in Africa by transitioning into water smart economies. This can be achieved through investment into new cost-effective technologies as well as enterprise innovations, which all aim to contribute to ensured water security. Decisive steps, however, need to be taken now as the country cannot afford to wait.

The book, therefore, contributes to and emphasises the general international consensus that water quality issues worldwide, specifically in South Africa, will have progressive and significant impacts or constraints on environmental and human health as well as economic development. This book ultimately achieves its main purpose by emphasising the fact that water is and will become an inescapable risk with the use of various case studies on a global scale as well as on a national South African scale with the establishment of water quality risk areas. The book finally calls for coordinated and decisive investment and action into addressing the highlighted water quality challenges.

Multiple opportunities exist in the water sector. Water, however, needs to be placed at the forefront of the global and South African agenda for these opportunities to be recognised, invested in and to ultimately ensure future sustainability to which the world should strive for.

Johannesburg, South Africa   Anja du Plessis

Acknowledgements

The author would like to thank the Department of Water and Sanitation's Resource Quality Information Services Department for water quality data for the selected physical, chemical and microbiological water quality parameters over the relevant time period as well as relevant GIS data. The author would also like to thank South African National Botanical Institute for the 2013/2014 land cover dataset.

Contents

Part I Water as a Global Risk

1 Current and Future Water Availability 3
   1.1   Realisation of Water as an Important Resource 3
   1.2   Current Water Availability 5
   1.3   Global Water Use 7
       1.3.1   Agricultural Water Use 7
       1.3.2   Industrial Water Use 8
       1.3.3   Domestic Water Use 9
   1.4   Future Water Availability and Conclusions................. 9
   References 11

2 Current and Future Water Scarcity and Stress 13
   2.1   Current Water Scarcity and Stress 13
   2.2   Case Studies of Current Water Scarcity and/or Stress 16
       2.2.1   Desiccation of the Aral Sea and Lake Chad 16
       2.2.2   China: Plentiful but Unusable 18
       2.2.3   Sri Lanka: Water Scarcity Due to Drought or a Lack of Planning 20
       2.2.4   The Horn of Africa—Persistent Cyclical Drought..... 22
   2.3   Conclusions 24
   References 24

3 Climate Change: Current Drivers, Observations and Impacts on the Globe's Natural and Human Systems 27
   3.1   Brief Introduction to Climate Change..................... 27
   3.2   Observed Changes Within the Climatic System 29
       3.2.1   Atmosphere 30
       3.2.2   Ocean 30
       3.2.3   Cryosphere 31
       3.2.4   Sea Level................................... 32

	3.3	Primary Drivers	32
		3.3.1	Natural and Anthropogenic Radiative Forcings	33
		3.3.2	Human Activities Affecting Emission Drivers	34
	3.4	Attributions of Climate Change	35
		3.4.1	Natural and Human Influences	36
		3.4.2	Observed Impacts	37
	3.5	Impacts, Vulnerabilities and Risks to Natural and Managed Resources	39
		3.5.1	Terrestrial and Inland Water Systems	40
		3.5.2	Food Security and Food Production Systems	43
	3.6	Impacts and Vulnerabilities of Human Settlements, Health, Livelihoods and the Economy	44
		3.6.1	Urban Settlements	45
		3.6.2	Rural Settlements	47
		3.6.3	Human Health	48
		3.6.4	Human Livelihoods	50
		3.6.5	Economy	51
	References	53
4	**Climate Change and Freshwater Resources: Current Observations, Impacts, Vulnerabilities and Future Risks**	55
	4.1	Brief Background	55
	4.2	Main Drivers	56
	4.3	Observed Changes and Attributions	57
	4.4	Impacts and Projected Hydrological Changes	58
	4.5	Projected Impacts, Vulnerabilities and Risks	60
		4.5.1	Water Availability	60
		4.5.2	Impacts on Water Use Sectors	61
	4.6	Regional Outlooks and Case Studies	65
		4.6.1	Africa: Vulnerable Rivers and Economies	68
		4.6.2	The Future of the Okavango Delta Under Climate Change	70
		4.6.3	Future of the Mitano Basin, Uganda Under Climate Change	71
		4.6.4	Pacific Islands	73
		4.6.5	Portugal	74
	4.7	Conclusions and Future Adaptation	75
	References	76
5	**Primary Water Quality Challenges, Contaminants and the World's Dirtiest Places**	79
	5.1	Brief Background	79
	5.2	Water Quality Challenges	80
		5.2.1	Primary Causes of Pollution and Main Water Quality Problems	81

		5.2.2	Primary Water Quality Challenges and Contaminants	84
	5.3		World's Top Pollutants, Toxic Threats and Dirtiest Places	98
		5.3.1	Worst Toxic Pollution Problems and Threats	99
		5.3.2	Most Polluted Countries and Rivers	103
	5.4		Conclusions	111
	References			112
6	**Water as a Source of Conflict and Global Risk**			**115**
	6.1		Water as a Source of Conflict	115
	6.2		Role of Increased Water Stress in Creating Conflict	116
		6.2.1	Parched Western USA Region: Severe Drought, Water Rights and Societal Impacts	117
		6.2.2	Possible Future Water Conflicts in India: Persistent Drought and Continued Poor Water Management	120
	6.3		Water Stress and Civil Unrest or Military Conflict	121
		6.3.1	East Africa Water Wars and Prolonged Civil Unrest in Sudan, Darfur	122
		6.3.2	Syria: A Country Unravelled	123
		6.3.3	Yemen: Humanitarian Crisis and a Non-existent State	125
	6.4		Water Stress and Competing Interests	126
		6.4.1	The Brahmaputra River: Sinking of China and India Relations	127
		6.4.2	The Renaissance Dam and Nile River: Ethiopia and Egypt	128
		6.4.3	Ilisu Dam and the Tigris River: The Case of Turkey and Iraq	129
	6.5		Future Climate Change and Water Conflicts	132
	6.6		Water as a Global Risk	134
		6.6.1	Costs of Lack of Clean and Reliable Water	135
		6.6.2	Costs of Insufficient Supply and Future Water Scarcity	136
	6.7		Managing Water Tensions, Risks and Conclusions	138
	References			141

Part II Water as a Regional and National Risk

7	**Evaluation of Southern and South Africa's Freshwater Resources**			**147**
	7.1		Background to Southern Africa's Freshwater Resources	147
		7.1.1	Water Availability	148
		7.1.2	Challenges and Vulnerabilities	151
		7.1.3	Possible Conflicts or Cooperation with Future Water Stress/Scarcity	154

7.2	Background to South Africa's Water Availability	155
	7.2.1 Surface Water	156
	7.2.2 Groundwater	158
7.3	Overview of Challenges and Vulnerabilities: Current and Future	161
	7.3.1 Growing Water Quality Problems	162
	7.3.2 Water Losses and Increased Water Stress and Scarcity	165
References		169

8 Establishing South Africa's Current Water Quality Risk Areas ... 173
  8.1 Brief Background to South Africa's Water Management Areas ... 173
  8.2 Establishing South Africa's Water Quality Risk Areas ... 175
  References ... 181

9 Current Water Quality Risk Areas for Limpopo, Olifants and the Inkomati-Usuthu WMAs ... 183
  9.1 Limpopo WMA ... 183
    9.1.1 WMA Overview ... 183
    9.1.2 Risk Areas for Domestic Use ... 186
    9.1.3 Risk Areas for Aquatic Ecosystems ... 187
    9.1.4 Risk Areas for Irrigation ... 188
    9.1.5 Risk Areas for Industrial Use ... 189
    9.1.6 *Chlorophyll a* and *Faecal Coliform* Risk Areas ... 190
  9.2 Olifants WMA ... 193
    9.2.1 WMA Overview ... 193
    9.2.2 Risk Areas for Domestic Use ... 195
    9.2.3 Risk Areas for Aquatic Ecosystems ... 196
    9.2.4 Risk Areas for Irrigation Use ... 197
    9.2.5 Risk Areas for Industrial Use ... 198
    9.2.6 Risk Areas for *Chlorophyll a* and *Faecal Coliform* ... 199
  9.3 Inkomati-Usuthu WMA ... 202
    9.3.1 Overview of WMA ... 202
    9.3.2 Risk Areas for Domestic Use ... 205
    9.3.3 Risk Areas for Aquatic Ecosystems ... 206
    9.3.4 Risk Areas for Irrigation Use ... 207
    9.3.5 Risk Areas for Industrial Use ... 208
    9.3.6 Risk Areas for *Chlorophyll a* and *Faecal Coliform* ... 208
  9.4 Conclusions ... 211

10 Current Water Quality Risk Areas for Vaal, Pongola-Mtamvuna and Orange WMAs ... 213
10.1 Vaal WMA ... 214
10.1.1 WMA Overview ... 214
10.1.2 Risk Areas for Domestic Use ... 216
10.1.3 Risk Areas for Aquatic Ecosystems ... 217
10.1.4 Risk Areas for Irrigation Use ... 218
10.1.5 Risk Areas for Industrial Use ... 219
10.1.6 Risk Areas for *Chlorophyll a* and *Faecal Coliform* ... 220
10.2 Pongola-Mtamvuna WMA ... 224
10.2.1 WMA Overview ... 224
10.2.2 Risk Areas for Domestic Use ... 227
10.2.3 Risk Areas for Aquatic Ecosystems ... 228
10.2.4 Risk Areas for Irrigation Use ... 229
10.2.5 Risk Areas for Industrial Use ... 230
10.2.6 Risk Areas for *Chlorophyll a* and *Faecal Coliform* ... 231
10.3 Orange WMA ... 234
10.3.1 WMA Overview ... 234
10.3.2 Risk Areas for Domestic Use ... 237
10.3.3 Risk Areas for Aquatic Ecosystems ... 238
10.3.4 Risk Areas for Irrigation Use ... 238
10.3.5 Risk Areas for Industrial Use ... 239
10.3.6 Risk Areas for *Chlorophyll a* and *Faecal Coliform* ... 241
10.4 Conclusions ... 244

11 Current Water Quality Risk Areas for Berg-Olifants, Breede-Gouritz and Mzimvubu-Tsitsikamma WMAs ... 247
11.1 Berg-Olifants WMA ... 248
11.1.1 WMA Overview ... 248
11.1.2 Risk Areas for Domestic Use ... 250
11.1.3 Risk Areas for Aquatic Ecosystems ... 251
11.1.4 Risk Areas for Irrigation Use ... 253
11.1.5 Risk Areas for Industrial Use ... 254
11.1.6 Risk Areas for *Chlorophyll a* and *Faecal Coliform* ... 255
11.2 Breede-Gouritz WMA ... 257
11.2.1 WMA Overview ... 257
11.2.2 Risk Areas for Domestic Use ... 260
11.2.3 Risk Areas for Aquatic Ecosystems ... 261
11.2.4 Risk Areas for Irrigation Use ... 262
11.2.5 Risk Areas for Industrial Use ... 263
11.2.6 Risk Areas for *Chlorophyll a* and *Faecal Coliform* ... 264

	11.3	Mzimvubu-Tsitsikamma WMA	268
		11.3.1 WMA Overview	268
		11.3.2 Risk Areas for Domestic Use	271
		11.3.3 Risk Areas for Aquatic Ecosystems	272
		11.3.4 Risk Areas for Irrigation Use	273
		11.3.5 Risk Areas for Industrial Use	274
		11.3.6 Risk Areas for *Chlorophyll a* and *Faecal Coliform*	275
	11.4	Conclusions	277
12	**South Africa's Water Reality: Challenges, Solutions, Actions and a Way Forward**		**281**
	12.1	South Africa's Water Reality	281
		12.1.1 Persistent Eutrophication	282
		12.1.2 Spreading Salinisation	283
		12.1.3 Mine Water	284
		12.1.4 Drowning in Sewage	285
		12.1.5 Poor or Complete Lack of Management or Responsibility	287
		12.1.6 Running on Empty	289
	12.2	Ultimate Losses and Costs	289
	12.3	Possible Solutions and a Way Forward	291
		12.3.1 Political Will	291
		12.3.2 Cost-Effective Technologies	292
		12.3.3 Policy Improvement	293
		12.3.4 Future Actions	293
	12.4	Conclusions	294
	References		295
Appendix: DWS Water Quality Guidelines			**297**

Abbreviations

AMD	Acid Mine Drainage
CMA	Catchment Management Agency
DWS	Department of Water and Sanitation
GDP	Gross Domestic Product
GHGs	Greenhouse Gasses
IPCC	Intergovernmental Panel on Climate Change
MDG	Millennium Development Goals
NRW	Non-Revenue Water
PCB	Polychlorinated Biphenyls
POP	Persistent Organic Pollutant
SADC	Southern African Development Community
SDGs	Sustainable Development Goals
UN	United Nations
USA	United States of America
WC/WDM	Water Conservation/Water Demand Management
WDM	Water Demand Management
WMA	Water Management Area
WWTWs	WasteWater Treatment Works

List of Figures

Fig. 8.1	South Africa's nine WMAs (RSA 2016).	174
Fig. 8.2	Evaluated physical and chemical water quality parameter sample stations	176
Fig. 8.3	Evaluated *Chlorophyll a* sample stations.	176
Fig. 8.4	Evaluated *Faecal coliform* sample stations	177
Fig. 8.5	Phases in the research procedure and their main actions	179
Fig. 9.1	The main catchments of the Limpopo WMA	184
Fig. 9.2	Water quality sampling sites used and land cover of the Limpopo WMA	185
Fig. 9.3	Overall risk profile and significant risk areas for the Limpopo WMA (domestic use standards)	186
Fig. 9.4	Overall risk profile and significant risk areas for the Limpopo WMA (aquatic ecosystem standards)	187
Fig. 9.5	Overall risk profile and significant risk areas for the Limpopo WMA (irrigation standards)	188
Fig. 9.6	Overall risk profile and significant risk areas for the Limpopo WMA (industrial standards)	189
Fig. 9.7	Overall risk profile of *Chlorophyll a* for the Limpopo WMA (domestic use standards)	190
Fig. 9.8	Overall risk profile of *Chlorophyll a* for the Limpopo WMA (recreational use standards)	191
Fig. 9.9	Overall risk profile of *Faecal coliform* for the Limpopo WMA (domestic use standards)	192
Fig. 9.10	Overall risk profile of *Faecal coliform* for the Limpopo WMA (irrigation standards)	192
Fig. 9.11	Main catchments of the Olifants WMA	194
Fig. 9.12	Water quality sampling stations used and land cover of the Olifants WMA	195
Fig. 9.13	Overall risk profile and significant risk areas for the Olifants WMA (domestic use standards)	196

Fig. 9.14	Overall risk profile and significant risk areas for the Olifants WMA (aquatic ecosystem standards)	197
Fig. 9.15	Overall risk profile and significant risk areas for the Olifants WMA (irrigation standards)	198
Fig. 9.16	Overall risk profile and significant risk areas for the Olifants WMA (industrial use standards)	199
Fig. 9.17	Overall risk profile of *Chlorophyll a* for the Olifants WMA (domestic use standards)	200
Fig. 9.18	Overall risk profile of *Chlorophyll a* for the Olifants WMA (recreational use standards)	200
Fig. 9.19	Overall risk profile of *Faecal coliform* for the Olifants WMA (domestic use standards)	201
Fig. 9.20	Overall risk profile of *Faecal coliform* for the Olifants WMA (irrigation standards)	202
Fig. 9.21	Main catchments of the Inkomati-Usuthu WMA	203
Fig. 9.22	Water quality sampling sites used and land cover of the Inkomati-Usuthu WMA	204
Fig. 9.23	Overall risk profile and significant risk areas for the Inkomati-Usuthu WMA (domestic use standards)	205
Fig. 9.24	Overall risk profile and significant risk areas for the Inkomati-Usuthu WMA (aquatic ecosystem standards)	206
Fig. 9.25	Overall risk profile and significant risk areas for the Inkomati-Usuthu WMA (irrigation standards)	207
Fig. 9.26	Overall risk profile for the Inkomati-Usuthu WMA (industrial standards)	208
Fig. 9.27	Overall risk profile of *Chlorophyll a* for the Limpopo WMA (domestic use standards)	209
Fig. 9.28	Overall risk profile of *Faecal coliform* for the Limpopo WMA (domestic use standards)	210
Fig. 9.29	Overall risk profile of *Faecal coliform* for the Limpopo WMA (irrigation use standards)	210
Fig. 10.1	Main catchments of the Vaal WMA	214
Fig. 10.2	Water quality sampling sites used and land cover of the Vaal WMA	216
Fig. 10.3	Overall risk profile and significant risk areas for the Vaal WMA (domestic use standards)	217
Fig. 10.4	Overall risk profile and significant risk areas for the Vaal WMA (aquatic ecosystem standards)	218
Fig. 10.5	Overall risk profile and significant risk areas for the Vaal WMA (irrigation standards)	219
Fig. 10.6	Overall risk profile and significant risk areas for the Vaal WMA (industrial standards)	220
Fig. 10.7	Overall risk profile of *Chlorophyll a* for the Vaal WMA (domestic use standards)	221

Fig. 10.8	Overall risk profile of *Chlorophyll a* for the Vaal WMA (recreational use standards)............................	221
Fig. 10.9	Overall risk profile of *Faecal coliform* for the Vaal WMA (domestic use standards)................................	222
Fig. 10.10	Overall risk profile of *Faecal coliform* for the Vaal WMA (irrigation standards)..................................	223
Fig. 10.11	Main catchments of the Pongola-Mtamvuna WMA	225
Fig. 10.12	Water quality sampling sites used and land cover of the Pongola-Mtamvuna WMA	226
Fig. 10.13	Overall risk profile and significant risk areas for the Pongola-Mtamvuna WMA (domestic use standards).........	227
Fig. 10.14	Overall risk profile and significant risk areas for the Pongola-Mtamvuna WMA (aquatic ecosystem standards).....	228
Fig. 10.15	Overall risk profile and significant risk areas for the Pongola-Mtamvuna WMA (irrigation standards).............	229
Fig. 10.16	Overall risk profile and significant risk areas for the Pongola-Mtamvuna WMA (industrial standards).............	230
Fig. 10.17	Overall risk profile of *Chlorophyll a* for the Pongola-Mtamvuna WMA (domestic use standards).........	231
Fig. 10.18	Overall risk profile of *Chlorophyll a* for the Pongola-Mtamvuna WMA (recreational use standards).......	232
Fig. 10.19	Overall risk profile of *Faecal coliform* for the Pongola-Mtamvuna WMA (domestic use standards).........	233
Fig. 10.20	Overall risk profile of *Faecal coliform* for the Pongola-Mtamvuna WMA (irrigation standards).............	234
Fig. 10.21	Main catchments of the Orange WMA...................	235
Fig. 10.22	Water quality sampling sites used and land cover of the Orange WMA.......................................	237
Fig. 10.23	Overall risk profile and significant risk areas for the Orange WMA (domestic use standards)	238
Fig. 10.24	Overall risk profile and significant risk areas for the Orange WMA (aquatic ecosystem standards)	239
Fig. 10.25	Overall risk profile and significant risk areas for the Orange WMA (irrigation standards)	240
Fig. 10.26	Overall risk profile and significant risk areas for the Orange WMA (industrial standards)	240
Fig. 10.27	Overall risk profile of *Chlorophyll a* for the Orange WMA (domestic use standards)...............................	241
Fig. 10.28	Overall risk profile of *Chlorophyll a* for the Orange WMA (recreational use standards)............................	242
Fig. 10.29	Overall risk profile of *Faecal coliform* for the Orange WMA (domestic use standards)...............................	243
Fig. 10.30	Overall risk profile of *Faecal coliform* for the Orange WMA (irrigation standards).................................	244

Fig. 11.1	Main catchments of the Berg-Olifants WMA	249
Fig. 11.2	Water quality sampling sites used and land cover of the Berg-Olifants WMA	250
Fig. 11.3	Overall risk profile and significant risk areas for the Berg-Olifants WMA (domestic use standards)	251
Fig. 11.4	Overall risk profile and significant risk areas for the Berg-Olifants WMA (aquatic ecosystem standards)	252
Fig. 11.5	Overall risk profile and significant risk areas for the Berg-Olifants WMA (irrigation standards)	253
Fig. 11.6	Overall risk profile and significant risk areas for the Berg-Olifants WMA (industrial standards)	254
Fig. 11.7	Overall risk profile of *Chlorophyll a* for the Berg-Olifants WMA (domestic use standards)	255
Fig. 11.8	Overall risk profile of *Chlorophyll a* for the Berg-Olifants WMA (recreational use standards)	256
Fig. 11.9	Overall risk profile of *Faecal coliform* for the Berg-Olifants WMA (domestic use standards)	257
Fig. 11.10	Overall risk profile of *Faecal coliform* for the Berg-Olifants WMA (irrigation standards)	258
Fig. 11.11	Main catchments of the Breede-Gouritz WMA	258
Fig. 11.12	Water quality sampling sites used and land cover of the Breede-Gouritz WMA	260
Fig. 11.13	Overall risk profile and significant risk areas for the Breede-Gouritz WMA (domestic use standards)	261
Fig. 11.14	Overall risk profile and significant risk areas for the Breede-Gouritz WMA (aquatic ecosystem standards)	262
Fig. 11.15	Overall risk profile and significant risk areas for the Breede-Gouritz WMA (irrigation standards)	263
Fig. 11.16	Overall risk profile and significant risk areas for the Breede-Gouritz WMA (industrial standards)	264
Fig. 11.17	Overall risk profile of *Chlorophyll a* for the Breede-Gouritz WMA (domestic use standards)	265
Fig. 11.18	Overall risk profile of *Chlorophyll a* for the Breede-Gouritz WMA (recreational use standards)	266
Fig. 11.19	Overall risk profile of *Faecal coliform* for the Breede-Gouritz WMA (domestic use standards)	267
Fig. 11.20	Overall risk profile of *Faecal coliform* for the Breede-Gouritz WMA (irrigation standards)	268
Fig. 11.21	The main catchments of the Mzimvubu-Tsitsikamma WMA	269
Fig. 11.22	Water quality sampling sites used and land cover of the Mzimvubu-Tsitsikamma WMA	270
Fig. 11.23	Overall risk profile and significant risk areas for the Mzimvubu-Tsitsikamma WMA (domestic use standards)	271

Fig. 11.24	Overall risk profile and significant risk areas for the Mzimvubu-Tsitsikamma WMA (aquatic ecosystem standards)	272
Fig. 11.25	Overall risk profile and significant risk areas for the Mzimvubu-Tsitsikamma WMA (irrigation standards)	273
Fig. 11.26	Overall risk profile and significant risk areas for the Mzimvubu-Tsitsikamma WMA (industrial standards)	274
Fig. 11.27	Overall risk profile of *Chlorophyll a* for the Mzimvubu-Tsitsikamma WMA (domestic use standards)	275
Fig. 11.28	Overall risk profile of *Chlorophyll a* for the Mzimvubu-Tsitsikamma WMA (recreational use standards)	276
Fig. 11.29	Overall risk profile of *Faecal coliform* for the Mzimvubu-Tsitsikamma WMA (domestic use standards)	277
Fig. 11.30	Overall risk profile of *Faecal coliform* for the Mzimvubu-Tsitsikamma WMA (irrigation standards)	278

Part I
Water as a Global Risk

Chapter 1
Current and Future Water Availability

Water has been widely recognised as a resource of vital importance due to the pivotal role it plays in human health, development and overall well-being. It is a fundamental human right and international goals have consequently been developed to achieve improved access and sustainability of water resources across the globe. The newly adopted Sustainable Development Goals aims to encourage countries around the world to achieve these set global goals by 2030.

Even though water is one of the most widely distributed substances, it is not always suitable for human consumption. Accessible and useable freshwater sources for human consumption is limited and unevenly distributed. The human populations do not always coincide with water-rich regions. It may be widely distributed and available, however, it is not always available when and where human populations need it. Different water uses affect water availability in regions. Main water use sectors use available for specific activities and may consequently negatively affect water sources through over-withdrawal or pollution through depositing wastes.

The combination of a constant increase in the world's human population as well as continued desire for economic growth will lead to intensified competition between the different water-usage sectors and place more pressure on the world's water resources as these sectors attempt to satisfy demands. A "Business as Usual" mentality is therefore becoming unsustainable and new "business unusual" strategies need to be developed.

1.1 Realisation of Water as an Important Resource

The effective and efficient management of water on a global and national scale has been deemed to be of vital importance due to it being crucial for various aspects of human health, development and well-being as well as a fundamental human right internationally.

The importance of this resource has once again been recognised in the newly adopted Sustainable Development Goals (SDGs). The SDGs or otherwise known as the Global Goals, were adopted by countries in 2015 as part of the new sustainable development agenda which will target the next 15 years. A total of 17 goals have been developed which focusses mainly on ending poverty, protecting the planet and ultimately ensuring prosperity for all. These 17 goals build on the Millennium Development Goals and include new areas such as climate change, economic inequality, innovation, sustainable consumption as well as peace and justice. These goals have been designed to be interconnected, meaning that a success on one goal will involve addressing issues more commonly associated with another. Clear guidelines and targets have been set for countries to adopt in accord with their own unique priorities and the environmental challenges across the globe (UNDP 2017; WHO 2018).

The world's water has been recognised as a priority and subsequently, a goal has been developed which focusses on ensuring access to water and sanitation for all. The clean water and sanitation goal distinguishes water as an essential part of the world we want to live in and recognises that sufficient freshwater is needed to achieve this. This goal tries to address the degradation of freshwater resources through unsustainable economics and poor infrastructure, to name but a few, which have led to millions of people, mostly children, die from water-related diseases accompanied with inadequate water supply, sanitation and hygiene. The following targets have been set to be achieved by 2030:

- Universal and equitable access to safe and affordable drinking water for all;
- Access to adequate and equitable sanitation and hygiene for all. Put an end to open defecation by paying special attention to the needs of women and girls and those in vulnerable situations;
- Improve water quality through the reduction of pollution, eliminating dumping and minimizing the release of hazardous chemicals and materials, halving the proportion of untreated wastewater and substantially increasing recycling and safe reuse globally;
- Increase water-use efficiency across all sectors and ensure sustainable withdrawals and supply of freshwater to address water scarcity and substantially reduce the number of people suffering from water scarcity;
- Implement integrated water resources management at all levels, including through transboundary cooperation where appropriate;
- Protect and restore water-related ecosystems, including mountains, forests, wetlands, rivers, aquifers and lakes;
- Expand international cooperation and capacity-building support to developing countries in water- and sanitation-related activities and programmes, including water harvesting, desalination, water efficiency, wastewater treatment, recycling and reuse technologies; and
- Lastly, support and strengthen the participation of local communities in improving water and sanitation management (UNDP 2017).

1.1 Realisation of Water as an Important Resource

The goal, therefore, emphasises that water scarcity, poor water quality as well as inadequate sanitation have a significant negative impact especially on food security, livelihood choices and educational opportunities for poor families around the world.

Consequently, other proposed goals have also been included related to water and sanitation targets and includes the following interconnected goals for 2030:

- *Goal 3.3*: End the epidemics of AIDS, tuberculosis, malaria and neglected tropical diseases as well as combat hepatitis, waterborne diseases and other communicable diseases;
- *Goal 3.9*: Substantially reduce the number of deaths and illnesses from hazardous chemicals and air, water and soil pollution and contamination;
- *Goal 11.5*: Significantly reduce the number of deaths and the number of affected people and decrease economic losses relative to Gross Domestic Product (GDP) caused by disasters, including water-related disasters, with the focus on protecting the poor and people in vulnerable situations;
- *Goal 12.4*: Achieve environmentally sound management of chemicals and all wastes throughout their life cycle in accordance with agreed international frameworks and significantly reduce their release to air, water and soil to minimize their adverse impacts on human health and the environment;
- *Goal 15.1*: Ensure conservation, restoration and sustainable use of terrestrial and inland freshwater ecosystems and their services, in particular forests, wetlands, mountains and drylands, in line with obligations under international agreements; and
- *Goal 15.8*: Introduce measures to prevent the introduction and significantly reduce the impact of invasive alien species on land and water ecosystems, and control or eradicate the priority species (UNDP 2017).

It is estimated that by 2050, at least one in four people is likely to live in a country which is affected by chronic or recurring shortages of freshwater. Water is, therefore, becoming an ever-increasing inescapable risk to all spheres of the world as the resource is continually being degraded and overused, causing widespread environmental degradation, water-related human health issues and increasing economic risk in terms of investment and future growth.

Part 1 of the book will, therefore, focus upon current water availability, distribution and use of freshwater resources, possible influences of climate change, water scarcity and stress as well as water-related problems, a possible source of conflict and as a global risk.

1.2 Current Water Availability

Water is one of the most widely distributed substances to be found in the natural environment and constitutes the earth's oceans, seas, lakes, rivers and underground water sources. However, not all of these widely distributed resources are suitable or available for human consumption.

Even though approximately 75% of the earth's surface is covered with water, only 3% of this is freshwater—the rest is seawater and undrinkable. However, this is just an estimate as the dynamic nature and permanent motion of water makes it difficult to reliably assess the total water stock/store of the earth. Current estimates are that the earth's hydrosphere contains approximately 1,386 million km^3 of water but not all of these resources are potentially available to humans since freshwater is required by the agricultural sector, industries and domestic and recreational users (Kibona et al. 2009; Cassardo and Jones 2011; Lui et al. 2011).

Of the 3% of total available freshwater, only 0.5% is available—the other 2.5% is frozen, locked up in glaciers, ice caps and permanent snow cover in the polar regions. The available 0.5% of freshwater occurs in different forms namely aquifers, rainfall, natural lakes, reservoirs and rivers. Underground aquifers or groundwater accounts for the most freshwater with 10 million km^3 stored in this form. Since the 1950s, there has been a rapid expansion of groundwater exploitation resulting in supplying 50% of drinking water, 40% industrial water and 20% of irrigation water globally. Rainfall falling on land accounts for 119,000 km^3, after accounting for evaporation, and natural lakes a total of 91,000 km^3 of the world's freshwater resources. Furthermore, man-made storage facilities or reservoirs have increased the global storage capacity of the world since 1950s sevenfold and account for over 5,000 km^3. Lastly, it is estimated that 2,120 km^3 is found in rivers which are constantly replaced from rainfall and melting of snow and ice (WBCSD 2006).

Freshwater is also not evenly distributed across the globe. Less than 10 countries hold approximately 60% of the world's total available freshwater supply. These countries include Brazil, Russia, China, Canada, Indonesia, United States of America (USA), India, Colombia and the Democratic Republic of the Congo. It should be noted that local variations within these countries also exist and can be highly significant.

It is important to note that both the human population and water resources are unevenly distributed across the earth's surface resulting in some areas that are densely populated by human populations not necessarily coinciding with regions that are rich in water supplies.

The minimum basic water requirement for human health is 50 L per capita per day and the minimum amount of water required per capita for food is approximately 400,000 L per year, as estimated by the World Business Council for Sustainable Development. Regions such as the USA consume more than eight times that amount for human consumption and four times that amount per year for food production confirming that water resources are unevenly distributed across the world (Kibona et al. 2009; Pimentel et al. 2010; Cassardo and Jones 2011).

The minimum requirements set for basic water requirement for human health indicates that the total amount of water available on Earth is sufficient to provide for the whole population. However, most of the total freshwater is concentrated in specific regions, while other regions such as the Middle East and North Africa face a water deficit (Cassardo and Jones 2011). Therefore, one might say that the world might have enough water but it is not always available when and where the human population needs it. Other factors such as climate, normal seasonal variations,

1.3 Global Water Use

The main water use sectors around the world can be grouped as agriculture, industrial and domestic water use and includes the following main water uses:

- *Agricultural water use*: irrigation, for livestock, for fisheries and for aquaculture,
- *Industrial water use*: economic entities such as mines, oil refineries, manufacturing plants, as well as energy installations using water for the cooling of power plants; and
- *Domestic water use*: drinking water, bathing, cooking, sanitation and gardening activities.

The competition for water resources between these water sectors are steadily increasing mainly due to the continued growth of the human population and aspirations of economic growth. On a global scale, agriculture accounts for 70% of water usage, followed by industrial use (22%) and domestic use (8%). Importantly, water usage is not uniform across regions and differs quite significantly between high-income and low-middle income countries. High-income countries are characterized by industrial usage being the most dominant water use sector, using 59%, followed by agricultural use (30%) and domestic water use (11%). Low- and middle income countries are dominated by agricultural water use accounting for 82%, followed by industrial use (10%) and domestic use (8%) (WBCSD 2006).

1.3.1 Agricultural Water Use

The agricultural water use sector dominates water use within developing nations where irrigation accounts for over 90% of water withdrawals. In comparison, regions where rain is abundant such as the case of England, agricultural water use only accounts for 1% whereas agricultural water usage exceeds 70% of total usage in Portugal and Greece which are located on the same continent. Agricultural water use, therefore, varies significantly according to the region/ country's characteristics.

The Green Revolution has been characterized by irrigation as the key component, which enabled many developing countries to produce sufficient food for their growing human populations. The sector will demand more water to produce more food for the increasing human population; however, this will increase the competition for water between water use sectors as well as inefficient irrigation practices which could constrain future production.

It is estimated that 15–35% of irrigation withdrawals are unsustainable globally which is a major concern for future food production and more regions are moving towards a high to moderate overdraft of irrigation withdrawals in an attempt to produce more food.

1.3.2 Industrial Water Use

Industrial water use is the second largest user of water globally and varies significantly from one type of industry to another. Water is used for energy in terms of hydro projects, cooling water for thermal power generation, process water, water for products and in some cases for waste disposal to name but a few. In this sector, business is largely dependent on water meaning that no water means no business.

The increase of competition between water sectors have caused some industries to look at alternative processes to reduce water used. Some examples include the following:

- A paper mill in Finland decreased the water used per unit by over 90% over the past 20 years through changing from chemical to thermo-mechanical pulp as well as the installation of biological wastewater treatment facility that enables the recycling of water;
- An Indian textile firm reduced its water use by over 80% by replacing zinc with aluminium in its synthetic fibre production. The firm reduced the trace metals in wastewater and consequently enabled reuse and the use of treated water for irrigation by local farmers; and
- A sugar production plant in Mexico decreased its consumption by over 90% through improving housekeeping and segregation sewage from process wastewater.

The largest single use of water is for the cooling in thermal power generation. In terms of process water, the industry uses water to produce steam for direct drive power and for use in various production processes and chemical reactions. Other industries which use water for products include businesses like beverage and pharmaceutical sectors, consume water by using it as an ingredient in finished products for human consumption (products delivered in liquid form). The term "virtual water" is consequently used to describe the water that is embedded in both agricultural and manufactured products as well as the water used in the growing or manufacturing process (WBCSD 2006).

Water is also used as a disposal site for wastewater or cleaning water into natural freshwater systems. Rivers and other water bodies can handle small quantities of waste broken down by nature, however, these limits are often exceeded and cause water quality to decline downstream causing it to be no longer useable without expensive treatment.

Industrial water usage is directly proportional to the average income level of its people. Industrial water withdrawals constitute 5% in low-income countries as

1.3 Global Water Use

opposed to the above 40% in some high-income countries. Furthermore, a number of countries in Asia are now also developing their economies around industrial development so that water usage in this sector will increase over subsequent years (Kibona et al. 2009; Lui et al. 2011). Industrial water use will, therefore, increase further across the globe, which may be accompanied with an increase of water-related risks such as increased competition for the resource as well as increased pollution if wastewater is continually disposed of into surrounding water bodies.

1.3.3 Domestic Water Use

Clean water and sanitation is an international human right which needs to be adhered to in all countries. Individuals, therefore, must have clean water for drinking and freshwater for cooking, washing and sanitation. The water use per capita for domestic use varies significantly across the globe. More than one billion people, mostly located in Asia, are still without improved drinking water and approximately 3,900 children die annually due to polluted water and poor hygiene. Approximately, 2.6 billion people are still without improved sanitation and the human population numbers are still increasing. The global coverage is mostly good in most regions across the globe.

The WHO/UNICEF Joint Monitoring Programme reported in 2004 that the world is on track to meet the drinking water target, however, the sub-Saharan Africa region is still lagging behind and the progress towards the sanitation target was too slow to meet the goal.

The access to water issue is directly related to distribution issues. A good example of this is that people living in urban slums are often close to water resources but infrastructure is not in place to supply them with clean drinking water and sanitation.

1.4 Future Water Availability and Conclusions

Water use sectors are facing increased rivalry between themselves in terms of water rights and withdrawals. Industrial water usage increases with the country's income and may consequently create increased difficulties in the allocation of water between the major water use sectors with increased economic growth. This may consequently become a significant business risk for industries as their increased need for water may not be able to be accommodated with the other water use sectors also demanding more.

The continuous growth in the human population is demanding an increase of agricultural outputs, which requires an increase of both water and energy consumption around the world. The primary challenge for this sector is to make 70% more food available in the next 40 years for the world's human population, which will increase the sector's water demand. Other water use sectors are facing the same predicament. It is predicted that global energy consumption will increase by 49%

from 2007 to 2035 which will consequently place more pressure on water resources in an attempt to increase production. In terms of the domestic water use sector, the basic household water requirement is 50 L per person per day which excludes gardens (Gleick 2006; Kibona et al. 2009). This estimate is, however, exceeded in most countries. Overconsumption as well as the goal of improving access to clean water and sanitation will also place further immense pressure on the world's resources as the demand increases. Another water use sector which also needs to be considered is recreational water use which includes the type of water use associated with reservoirs and activities such as boating, angling, water skiing, as well as swimming, to name a few. Even though this water sector has a low-water consumption of only 1% of the world's water resources, the sector is slowly increasing and may have further consequences for other water use sectors such as agriculture by reducing the availability of water for users at specific times. A good example of this is in the event of water being retained in a reservoir to allow for boating in late winter. The storage of water in the specific reservoir may cause this water to be unavailable to farmers during the spring planting season (Kibona et al. 2009; Lui et al. 2011).

Lastly, environmental water usage uses the least water of the main water usage categories and benefits ecosystems rather than human beings. Notwithstanding this, the total water usage is increasing as a result of artificial wetlands and lakes that are intended for creating habitats for various wildlife species. As in the case of recreational water usage, environmental water usage is non-consumptive but may reduce the total volume of water that can be made available to other users at specific times and locations. With an increase in the adoption of ecocentric and biocentric value systems, we can expect more water to be directed in the future to ecosystems and nature reserves than to human needs (Kibona et al. 2009; Lui et al. 2011; UN 2012).

The combination of a constant increase in the world's human population as well as continued desire for economic growth will consequently lead to intensified competition between the different water-usage sectors and place more pressure on the world's water resources as these water use sectors attempt to satisfy demands. Physical evidence already suggests that human activities have already reached or even exceeded the renewable water limits in numerous regions across the globe. A clear indicator of unsustainable water usage is the now common practice of chronic over-extraction of groundwater in numerous important food-producing regions and in large urban areas. Much of China's North Plain, the USA's Great Plains and California's Central Valley, parts of the Middle East and North Africa, the valley of Mexico and parts of Southeast Asia are exceeding their groundwater recharge levels. Another example would be the lower reach of the Yellow River located in China which ran dry every year the past decade during all or part of the dry season when irrigation farming is at its most prolific—pointing to excessive water usage (Kibona et al. 2009).

A "Business as Usual" mentality is, therefore, not an option and new "business unusual" strategies need to be developed. These strategies need to be financially viable, socially acceptable and ecologically sustainable. To decrease or mitigate water-related risks for industries as well as ensure sustainable water usage, trade-offs will have to be made and water management strategies and technologies should

1.4 Future Water Availability and Conclusions

be developed or improved upon on all scales, i.e. global, regional and national/local. The increasing threat of water scarcity and stress around the world also needs to be taken into account as this will put added pressure on water availability as well as create increased competition between water sectors as each demand more with the continued increase in population and economic growth.

References

Cassardo C, Jones JAA (2011) Managing water in a changing world. Water 3:618–628

Gleick P (2006) The world's water: the biennial report on freshwater resources. Island Press, Washington

Kibona D, Kidulile G, Rwabukambara F (2009) Environment, climate warming and water management. Transit Stud Rev 16:484–500

Lui J, Dorjderem A, Fu J, Lei X, Lui H, Macer D, Qiao Q, Sun A, Tachiyama K, Yu L, Zheng Y (2011) Water ethics and water resource management. Ethics and climate change in Asia and the Pacific (ECCAP) project, working group 14 report. UNESCO, Bangkok

Pimentel D, Whitecraft M, Scott ZR, Zhao L, Satkiewicz P, Scott TJ, Phillips J, Szimak D, Singh G, Gonzalez DO, Moe TL (2010) Will limited land, water, and energy control human population numbers in the future? Hum Ecol 38:599–611

UN (United Nations) (2012) The United Nations world water development report 4, vol 1–3. UN

UNDP (United Nations Development Programme) (2017) Sustainable development goals. Available via http://www.undp.org/content/undp/en/home/sustainable-development-goals/background.html. Accessed 25 Feb 2018

WBCSD (World Business Council for Sustainable Development) (2006) Facts and trends: water. WBCSD, Earthprint Limited

WHO (World Health Organisation) (2018) Millennium development goals (MDGs). Available via http://www.who.int/topics/millennium_development_goals/about/en/ Accessed 25 Feb 2018

Chapter 2
Current and Future Water Scarcity and Stress

Current trends show that numerous water systems which sustain ecosystems, as well as growing human populations, have become stressed. Water sources are either being degraded by various types of pollution becoming unsuitable for use or have dried up. The different water use sectors have contributed to increased water scarcity by increased water wastage through inefficiencies as well as pollution. Climate change has also attributed to water shortages and droughts in some areas and floods in others around the globe. This chapter contains various case studies which emphasise the different manners in which water scarcity or stress can occur in various regions around the world.

Current consumption rates and further expansion and growth of the human population will exacerbate water stress and scarcity within regions and increase the rate of ecosystem degradation around the world. This should be a concern for all as it further increases water as a global risk.

2.1 Current Water Scarcity and Stress

The combination of increasing population growth, demographic changes, urbanisation, climate change and increasing water scarcity is posing numerous challenges for water supply systems. Water scarcity is defined as the lack of sufficient available water resources to meet a particular region's water needs. It has been widely documented that an increased number of countries around the world is entering an era of severe water shortages and limitations in freshwater supply are already evident where large areas are facing amplified consequences of dwindling and disappearing water reserves. Previously, most attention has been placed on Middle Eastern and North African countries, however, recently other regions have started to receive attention.

It is estimated that approximately four billion people worldwide are affected by severe water scarcity for at least one month a year which is much higher than the previous estimation of between two and three billion. This estimation was calculated by looking at people's water footprint from month to month and comparing this with the monthly availability of water in largest river basins. Particular problem areas which feel the effects of water scarcity directly were identified as Mexico, Western USA, Northern and Southern Africa, Southern Europe, Middle East, India, China as well as Australia farmers, industries and households which regularly experience water shortages. Other areas' water supplies were estimated to be fine but at a long-term risk (Mekonnen and Hoekstra 2016). Most of the sub-Saharan regions is characterised by economic water scarcity while more and more regions in developed countries such as the USA and China are approaching physical water scarcity.

Recent research has highlighted these alarming figures and has placed the water scarcity increasingly higher subject on the global agenda. With the human population estimated to expand to nine billion by 2050 accompanied by increased urbanisation, the demand for freshwater resources is set to increase dramatically. We therefore need to look at water stress within regions which occurs when water demand exceeds the available amount during a certain period or when poor water quality restricts use due to it being of an unsuitable standard for use.

Numerous countries around the world are experiencing different degrees of water stress. A country is deemed as water stressed when the annual per capita renewable freshwater availability is less than 1,700 m^3. Increased number of regions and countries around the world are entering a period of water stress. 60% of European cities with more than 100,000 people are using groundwater at a faster rate than it can be replenished. Other cities such as Mexico City, Bangkok, Beijing, Madras and Shanghai have experienced aquifer declines of between 10 and 50 m. The human population has succeeded in harnessing most of the world's natural waterways by building dams, water wells, irrigation systems as well as other infrastructure which have allowed the human population to expand. This expansion has, however, been accompanied with increased stress on these water systems and has led to some rivers, lakes and aquifers to run dry. The main causes or threats to increased water stress around the globe, but not limited to, include water pollution, inefficient water use especially in the agricultural sector, continued population growth as well as climate change (WWF 2017).

Water Pollution Water systems can be polluted by various point and non-point sources which can include untreated human and industrial wastewater and waste as well as pesticides and fertilisers washing into systems from farms. These effects can be immediate or take a period of time to build up in the environment and food chain before the effects are recognised. A section within this chapter will elaborate more on the main water quality challenges which exacerbate water stress or scarcity within the different regions of the world.

Inefficient Agricultural Water Use The agricultural sector is the biggest water user in the world, however, 60% of its water use is deemed as being wasteful or inefficient due to leaky irrigation systems, inefficient application methods as well as planting "thirsty" crops. This sector has led to the drying out of some rivers, lakes and aquifers and big food producing countries such as India, China, Australia, Spain and USA are facing increased water stress by having reached their water resource limits. The combination of agriculture's high water demand, as well as extensive water pollution through fertilisers and pesticides, have placed this water use sector as one of the main threats affecting both humans and other species.

Population Growth The doubling of the human population in the last five decades with the accompaniment of economic development and industrialisation has led to the transformation of the world's ecosystems and massive loss of biodiversity. Approximately, 41% of the world's population lives in river basins that are under water stress and the concern about water availability is increasing as freshwater use continues to be unsustainable. The continual increase of the human population requires continual increase for the demand for food, shelter and other resources which will result in additional pressure on freshwater resources. The increased demand for water will be driven primarily by the growing demand for food which is expected to increase by 70% in 2050 causing agricultural water use to increase by at least 19%. The increase of consumption will also be accompanied with increased water demand and use in manufacturing and production sectors.

The combination of the continued growth in the human population, as well as increased urbanisation, is creating further challenges and risks in terms of water scarcity and stress. The pressure that the world's human population places on available water resources varies across the globe. It is estimated that the global urban population will reach 66% by 2050 and is accompanied by major concerns of overexploitation and pollution of water resources. Developing regions such as Africa and Asia are the two regions already suffering the most from lack of urban water supply and sanitation. An example of this is that 22 out of 32 major Indian cities deal with daily water shortages. The main driving forces, therefore, include a steady increase in water demand, inefficiencies in transfer and use as well as persistent water pollution. The lack of adequate infrastructure such as proper wastewater treatment and drainage facilities are leading to continued pollution of ground and surface water supplies.

Numerous scientists and academics have highlighted the pending predicament for all humankind that will result from high population growth rates and limited natural resources, the latter having been exploited and been depleted since the 1800s. Even though this physical end has not yet been reached, a crisis point from an international security standpoint is already evident in many regions. A large proportion of the world's population will be confronted with severe risks that will be triggered by inevitable droughts in their regions. The water shortages are expected to multiply in the future—even in the absence of the factor of global warming or climate change—as a result of mismanagement and pollution (Shen et al. 2008; Frederiksen 2009). Unless the human population learns to cope with less water, more and more communities will face water scarcity and their regions running dry.

Climate Change The continued effects of climate change on the world's resources are increasingly being recognised or noticed around the world. The continual pumping of carbon dioxide as well as other greenhouse gases into the atmosphere have been accompanied with changes in weather and water patterns around the world and have consequently led to increased occurrences of droughts, floods and other extreme weather events. These changes can make less water available for water use sectors around the world and increase water stress. Chapter 4 investigates the possible influence of climate change on the world's freshwater resources.

It is important to note that the World Economic Forum has placed the world water crisis in the top three of global problems alongside climate change and terrorism. This threat is quite clear as numerous regions across the globe are faced with water stress. The sub-Saharan African region has the largest number of water-stressed countries of any region in the world.

More than 80 nations across the globe are characterised by water demands exceeding available supplies, more than 300 cities across China have inadequate water supplies and in some arid regions such as the Middle East and parts of North Africa, the low annual rainfall and expensive irrigation techniques have culminated in a grim scenario for agriculture in the future.

Impacts or consequences of increased water scarcity and stress can be severe and have everlasting effects on a region. Some of these impacts include the disappearing of wetlands, damaged ecosystems as well as the lack of water for billions of people around the world. Clean freshwater is essential for the functioning of ecosystems as well as for human health. When waters run dry, people are unable to access enough water to drink or feed crops and can lead to economic decline. Furthermore, inadequate sanitation may also lead to increased water-related diseases such as deadly diarrheal diseases which include cholera, typhoid fever and other waterborne illnesses.

2.2 Case Studies of Current Water Scarcity and/or Stress

2.2.1 Desiccation of the Aral Sea and Lake Chad

The cases of the disappearance of the Aral Sea located in Central Asia as well as Lake Chad in Central Africa are perfect examples of the consequences of excessive surface water withdrawal causing water stress within a region.

The Aral Sea was the fourth largest saline lake in the world up to the twentieth century and is fed by two rivers namely the Amu Darya and the Syr Darya located, respectively, in the South and North. These rivers were, however, diverted by the Soviet Union in the 1960s to irrigate the desert region surrounding the sea for agricultural purposes rather than feed the Aral Sea basin. The construction of these canals was aimed at developing a massive industry on cotton farming but led to dramatic negative consequences for the basin.

The Aral Sea was deprived of its main sources of water income and it was estimated that the majority of the water that was being diverted through canals were soaked up by the desert and blatantly wasted (25–75% season dependent). The diversion of the rivers caused an imbalance in the water system and led to the slow desiccation of the sea since the 1960s.

Lake Chad, located on the southern border of the Sahara, was an attractive oasis for four separate nations which laid claim to it as their borders intersect in the middle of the body of water. The basin supports more than 20 million people with the upper part of the catchment supporting fishing, agriculture and pastoralism. The increase in the surrounding region's population led to an increase in agricultural water demand for irrigation (Jacobs 2007). This with the accompaniment of warmer summers led to excessive water use within the region and led to Lake Chad being reduced to one-twentieth of its original size. Overgrazing, as well as unsustainable irrigation, resulted in the replacement of natural vegetation with invasive plant species, deforestation and the drying of the climate.

The consequences for the Aral Sea region has included hotter summers and colder winters, dust storms as well as "salt storms" which has swept across the region and bombards the surrounding environment with sand and toxic chemicals which have been left behind by Soviet factories built on once islands isolated from the sea's shore. Two small portions remain in the north and south. The Northern portion may survive as extensive work is being done to make sure that it does not receive the same fate as the rest of the region, however, the Southern part is drying up year after year and has no inflow from any rivers. The main environmental impact has therefore been the significant loss of water accompanied by a sharp increase of salinity exceeding the threshold of commercial fish.

Other impacts included a change in sea surface temperature (hotter in summer, colder in winter), desertification of the surrounding region, salinization of soil, and climate change. These environmental effects have caused the population growth rate in the Aral Basin to become diminished. Other regions have experienced an increase in population as well as water demand and have forced Uzbekistan to trade water for natural gas with Kyrgyzstan. The idea of the Soviet Union did, however, work in the sense that today cotton is Uzbekistan's primary export. However, the environmental price that has been paid is that the Aral Sea is now massive expanses of salt plains and a dry, barren graveyard surrounded by dozens of ghost towns which thrived on the fishing industry (Ataniyazova 2003; White 2013).

The consequences for the Lake Chad region may include future wars and civil violence as water scarcity intensifies with climate change and the lifeblood of Lake Chad's 20 million beneficiaries decreases with continued population growth, spread of disease, oppression and corruption. The fluctuations of the lake's water level have also caused considerable changes in fish fauna leading to high mortality, disappearance of some open water fish species as well as appearance of species adapted to swamp conditions which were previously unknown. The surrounding population is consequently relying less on fishing and shifting more towards the raising of cattle, sheep and camels as well as the farming of the emergent lake floor as flood waters recede. The region is also threatened by desertification due to climatic irregularity, occasional extreme droughts, unsustainable management of natural resources and population increases throughout the region. Other impacts also include the spread of invasive alien grasses and improper water management which include the building of dykes and lack of proper irrigation systems have led to the accumulation of salt in the soil (FAO 2017).

Therefore, the disappearance of the Aral Sea and Lake Chad can be seen as prime examples of how excessive water use within a region and poor water planning and management could lead to increased water scarcity within a region and have severe consequences for the environment as well as the surrounding population. There are numerous other examples of disappearing water resources over the world which are also attributed to excessive water use. We must therefore highlight these issues and stay mindful of that excessive water use accompanied with a lack of proper water planning and management can have severe unintended consequences.

2.2.2 China: Plentiful but Unusable

The dominant view or image associated with water scarcity is primarily the excessive overuse of water resources leading to dry riverbeds and lakes, however, this is not the only factor which plays a role as the case of China will illustrate. Until recently, most focus has been placed on the country's immense air pollution, however, it is now stated that water pollution is as serious an issue.

China contains approximately 20% if the world's population, however, only 7% of the world's available water resources. The country has enjoyed double-digit economic growth over the past three decades making it one of the world's largest economies but not without a price. The region has been facing increased water scarcity, particularly within the northern part of the country which has recently started to spread to the south. The increased water scarcity and stress is mainly attributed to insufficient local water resources as well as improper water quality due to continued increased pollution. The uneven spatial distribution of water resources within the country, the continued rapid economic growth and urbanisation of a large and growing population with poor resource management have caused China to be very vulnerable in terms of its water resources (Jiang 2009; Tiezzi 2014; Jing 2016).

Water scarcity has become more prominent within the region as it has spread from the north and west to the south and east with major lakes along the Yangtze River drying up. This situation is further exacerbated with intense pollution causing water to be unusable. Parts of the Poyang Lake, the largest freshwater lake in the country, have been drying up and become grassland within the winter months due to overexploitation of water resources upstream. A similar situation is present at the Dongting Lake, located in Hunan, where inflows from the Yangtze River have decreased by 40%. Rivers in the northern regions of the country have also been overexploited beyond international safety limits. Sustainable groundwater extraction limits have also been surpassed by extracting 6 billion m^3 for industrial and agricultural growth and has consequently led to villagers in Wei county having to dig 360 instead of 120 m deep to reach groundwater resources (Economy 2013; Shemie and Vigerstol 2016).

The country's water resources are further plagued by excessive industrial wastewaters and pollution as well as fertilisers and pesticides from the growing agricultural sector which has made many resources unusable. Excessive pollution is causing the more plentiful water resources located in the south and east of the country to become scarcer and have consequently expanded the country's water crisis. Water pollution and shortages is a major problem in Northern China with 45% estimated to be unfit for consumption compared to 10% in the south. The northern province of Shanxi has been affected the worst with 80% of rivers being rated unfit for human contact. The situation will become even more severe as the country's economy continues to grow.

The country is, therefore, experiencing a severe water pollution crisis with nearly 80% of the groundwater being polluted and is unsafe for drinking, up to 40% of rivers are seriously polluted and 20% are polluted to such an extent that their water quality is rated as being too toxic to come into contact with. Main causes for water pollution include wastewater from factories such as the 10,000 photochemical plants located along the Yangtze and 4,000 along the Yellow Rivers as well as sewerage pollution from more than 80% of treatment facilities not meeting government standards. China has focussed predominantly on industrialisation and economic growth which has been accompanied with widespread destruction of the natural environment and ecosystems through smog, sewage and industrial wastewaters and litter (Economy 2013; Jing 2016; Shemie and Vigerstol 2016).

Approximately one third of industrial wastewater and more than 90% of household sewage is estimated to be released untreated into rivers and lakes around the country. This is mainly attributed to almost 80% of cities not having proper or any sewage treatment facilities. Half of the country's population lacks safe drinking water and an estimated more than 500 million people in rural areas use water contaminated by human and industrial waste. Lakes are often affected by pollution-induced algal blooms and multiple rivers are unsuitable for human contact. The pollution and waste associated with the agricultural sector did not receive enough attention and led to pollution being double the predicted estimate set by the government in 2010 and a pollution census revealed that fertiliser is a bigger source of water contamination than factory effluent.

The effects of this widespread pollution have been detrimental towards the environment and human health. Some effects include mass mortality of fish as rivers are covered with layers of floating litter and contaminated by agricultural, industrial and human waste. Deformities in fish have also been recorded and been blamed on a paint chemical widely used in the industry. The pollution is also spreading and China is now described as the largest polluter of the Pacific Ocean with the development of offshore dead zones and red tides. Pollution has also affected the Chinese economy by causing economic losses of approximately US$69 billion per year. Human health has also been seriously affected with gastrointestinal cancer now being the number one killer in the rural areas due to two-thirds of the rural population having to use water contaminated by human and industrial waste. Some village residents within the Guangxi Province in Southern China were poisoned by arsenic-contaminated water which has been attributed to the waste from a metallurgy factory located nearby. Some villages are now being described as "Cancer Villages" where cancer rates have risen dramatically due to pollution. There are said to be around 100 cancer villages along the Huai River and its tributaries in Henan Province, especially on the Shaying River. Death rates on Huai River are 30% higher than the national average. Other health effects include diarrhoea, bladder and stomach cancer, other diseases caused by waterborne pollution, miscarriages and birth defects as well as death (Jing 2016; Shemie and Vigerstol 2016).

The country has acknowledged the urgency to address the water crisis and have provided funding to local governments to address water issues from water conservation and water diversion projects primarily aimed at "stabilising economic growth" by raising employment and investment levels. Focus has also been placed on increasing water efficiency and decreasing water use and has worked on the US$60 billion South–North Water Diversion Project which will bring water from the moisture-rich south to the dry north. The acknowledgement of the water crisis, as well as the development of various pollution prevention work, are encouraging however it was found in 2016 that US$2.56 billion was not effectively used and seems like corruption may hinder the needed progress (Jing 2016; Shemie and Vigerstol 2016).

Therefore, the degradation of water resources is as big as a threat to increase water scarcity within a region as excessive water use. The combination of these two factors can be detrimental to both the environment as well as the region's population as shown in this case study. Countries, therefore, need to take note of this and highlight the importance of proper water treatment infrastructure or technologies as well as enforcement of regulations to ensure that wastewater is treated properly to decrease the further degradation of water within regions. The combination of excessive water use as well as the continued degradation of water quality can be detrimental to a region and lead to increased water scarcity if proper steps and measures are not taken to address the various identified issues.

2.2.3 Sri Lanka: Water Scarcity Due to Drought or a Lack of Planning

Sri Lanka, a developing country, has been described as a country with untapped surface and groundwater resources with water resource potential, estimation of some major competition for water resources between sectors, an unlikelihood of severe water shortages with water quality as well as excessive extraction of groundwater becoming an issue of concern. The country has an average rainfall of 2,200 mm per annum with drier regions having 900 mm per annum. The agricultural sector accounts for 85% of the water demand through irrigation. The country is heavily dependent on both rain-fed and irrigated agriculture and the sector forms the pillar of rural livelihoods (IWMI 2007).

It has been estimated that the country is not likely to face a water crisis situation by the year 2025 but will have some areas of concerns in terms of quality and groundwater extraction which needs to be monitored and properly managed. Some research has suggested that the overall rainfall of the country has decreased in some areas, that rainfall patterns have changed and that the distribution of rainfall in certain parts of the country have also been undergoing changes. Other research also indicates that aggregated national level statistics on water scarcity has been misleading as recent statistics indicate that water availability will be a significant constraint in certain districts and will indeed be a significant constraint on social and economic development (IWMI 2007; Rodrigo and Senaratne 2013).

The country has also experienced an increase in water demand by sectors such as hydropower, domestic and industry along with the population growth. Increased water demand within the various sectors will and may lead to conflict between water use sectors. Water quality within the country has also steadily decreased due to increased urban pollution, agrochemical pollution which will exacerbate water scarcity by leaving water resources unusable. As most other developing countries, Sri Lanka also has a high water inefficiency in many irrigation schemes which also needs to be taken into account. Water scarcity will therefore indeed be an issue in the country and requires proper planning and management to ensure that the country does not experience a water crisis in the near future (Rodrigo and Senaratne 2013; Fernandez 2017).

The country has recently been experiencing the worst drought in four decades and has fallen into a water crisis with more than a million people experiencing acute water shortages and with more shortages expected in the near future. The lack of rain associated with the drought has caused water resources to become saline and not suitable for drinking. Rain is not expected, water levels of reservoirs are declining to a fifth of their capacity and the agricultural and hydropower generation sectors have also been affected. The drought has caused a delay in the planting seasons and has been responsible for crop damages. Farmers have only managed to plant a third of the usual paddy fields, predictions show that half of the crops may fail and the next planting season is at risk. Several districts in the country have been affected by water shortages (Fernandez 2017).

Efforts have been made by the government to address some water-related issues by establishing new infrastructure, rehabilitating or renovating dams, reservoirs, canals as well as promoting agro-wells and micro-irrigation technologies to address the rising agricultural water demand. Despite these efforts, water scarcity problems have continued to rise.

The lack of policies, improper planning as well as the lack of public awareness related to the extent of the country's water problems has been identified to be the biggest culprit for the country's water shortage problem. The country has been experiencing more extreme weather conditions and has been alternating between drought and floods, both which have led to shortages of clean water. This pattern has been identified by experts and policymakers and effective national interventions have not been put in place to address the identified issues. An example of this is the reluctance of officials to release water in storage of the largest reservoir in the Polonnaruwa District that is currently at 50% capacity as they fear that there will not be enough water for irrigation for farmers to plant in the next growing season which starts in November. The water demand, as well as competition for water, is growing within the country and better cooperation is required between the various government sectors as currently very little coordination exists. The lack of public awareness related to the extent of the country's water problems has also contributed to the country's water supply problems. Residents within the Poddibanda village is a good example of this as they are at the mercy of changing rainfall patterns and have had little assistance in learning to manage their water resources better. The improvement of water management within villages are important for their economic well-being as most farmers who are water short are preparing for cultivation as they expect the government to provide free irrigation water, regardless of rain or reservoir levels. Public awareness campaigns are therefore needed to build knowledge within these communities about water management practices and national water management policies are required to coordinate water planning between water sectors (Rodrigo and Senaratne 2013; Perera 2016).

The country will experience aggravated water scarcity as increasing evidence suggest that climate change will exacerbate water availability and quality issues. New innovative approaches are required in the agricultural sector to meet its future demands. Water balances also need to be achieved between the demands of the agricultural sector and the municipal and industrial water use sectors. The achievement of water balance between these major water use sectors will enable the country to carefully assess their set public health, environmental protection, economic viability as well as their food security goals. Other steps which can be taken are the development of crop varieties which have a lower water demand. It is therefore of prime importance that different government agencies such as the Department of Agriculture, Department of Irrigation as well as the Department of Meteorology recognise that they have an important role to play and that they have to work closely and share knowledge and information to improve water management strategies within the country (Rodrigo and Senaratne 2013; Perera 2016).

2.2.4 The Horn of Africa—Persistent Cyclical Drought

The Horn of Africa, which includes Kenya, Ethiopia, Somalia, Uganda and Djibouti, experienced the worst drought in 60 years in 2011 which threw millions of people across the region into one of the worst crises. Somalia which has been ravaged by conflict experienced the most disastrous consequences and hundreds of thousands of people, many of them children, have starved to death.

The United Nations Food and Agriculture Organisation has once again declared that the region is facing food shortages owing to the persistent lack of rainfall between October and December and as many as 12 million people, are in need of food assistance. This is, however, not a new occurrence within the Horn of Africa as the region has experienced persistent drought since 2014. The region has been caught up in this cycle which is likely to repeat itself despite countries attempting to prevent drought and other climate-related disasters. With little rain and none expected for the next couple of months, it has been warned that 17 million people are facing hunger in the region (FAO 2016).

The region is likely to see a rise in hunger as well as the future decline of local livelihoods as the population is struggling with the knock-on effects of the consequent droughts that it has been experiencing. The situation is further intensified by the rise in refugees in East Africa, placing additional pressure on the already strained food and nutrition security of the region. The Horn of Africa is experiencing severe water shortages and some families within Kenya have resorted to eating only one meal a day in an attempt to conserve fading food supplies. The region's environment has also been affected (Ngumbi 2017; ActionAid 2011).

Climate change has been identified by scientists as being the main factor which has been exacerbating the continued droughts within the region and it has been estimated that the situation will worsen as the increase of global temperatures have resulted in droughts becoming more frequent and severe and heavy rains causing more flooding. Climate experts assumed there would be an increase in rainfall in rainy seasons due to the rise in the Indian Ocean water temperature, however, recent trends indicate that the region is drying faster than anytime in the past 2,000 years.

Not only the human population have been affected by the prolonged droughts, ecosystems such as the Serengeti, have also been adversely affected. The prolonged droughts have impacted upon the grassland and its ability to sustain large herds of herbivores as well as the level of the Mara River and Lake Victoria which is the lifeblood of the Serengeti migration as well as a source of water and livelihood for large populations of Tanzania and one-third of the East Africa population. The flow of the Mara River has already been reduced by 25% and is increasingly being polluted by pesticides, sewage and phosphates. Flash floods which have followed periodic droughts have also eaten away vegetation on the river banks increasing soil erosion and sedimentation. The declining level of Lake Victoria is a critical issue. Tanzania uses the largest share of the lake, however, Uganda controls the outflow. Uganda has constructed two dams and has increased the discharge of the White Nile River by 30–50% which has led to the reduction of the lake's water level by two metres. Another dam is being constructed which has destroyed large areas of papyrus swamps and have placed the fishing industry under immense risk through the reduction of fish populations and breeding areas (ActionAid 2011; FAO 2016).

Rapid intervention is therefore required in terms of the affected populations as well as areas constantly affected by natural hazards to assist in their ability to withstand or adapt to the impacts accompanied by these persistent droughts. Predictions indicate that some regions such as Southern Kenya will be less affected by the drought mid-2017; however, regions in the north may worsen. As indicated previously, some strategies have been implemented in the region. Short-term strategies include the distribution of food and long-term strategies include the planting of drought-tolerant crops or diversifying crop and income base of communities. Kenya has also started to invest in community water sources in an attempt to decrease the dependence on rain-fed agriculture and have established a national drought management authority (ActionAid 2011; FAO 2016; Ngumbi 2017).

The Horn of Africa is gripped in persistent droughts and water scarcity which is a great threat to various countries, millions of people as well as the environment which contributes to the region's economy through ecotourism. Climate change will exacerbate the current situation through prolonging droughts as well as increasing extreme weather events which will increase hazardous areas in the region. The Horn of Africa region is, therefore, a prime example of how climate change can exacerbate water scarcity within a region through prolonging droughts and increasing hazardous events such as flash floods.

2.3 Conclusions

In conclusion, as indicated by previous sections and the included case studies, numerous water systems which sustain ecosystems, as well as growing human populations, have become stressed around the world. Rivers, lakes and aquifers are either being degraded by pollution to such an extent that it is unsuitable for use or are drying up. More than half of the world's wetlands have been destroyed, agriculture has the highest water use and contributes to increased water scarcity by increased water wastage through inefficiencies as well as pollution and lastly, climate change have caused water shortages and droughts in some areas and floods in others.

The current consumption rate as well as the human population estimated to be 9.6 billion in 2025, will increase water stress and scarcity within regions even more and increase the rate of ecosystem degradation around the world more. With climate change estimated to exacerbate water stress and scarcity around the world, it therefore needs to be acknowledged as a serious threat to the sustainability of the environment and human livelihoods as well as increasing competition between water use sectors, promote further water-related risks and possibly conflict. The following two chapters will, therefore, focus on the multiple facets of climate change and its predicted consequences to establish which regions will be more at risk of increased water stress, scarcity and other water-related risks.

References

ActionAid (2011) East Africa drought questions and answers. Available via http://www.actionaid.org/what-we-do/emergencies-conflict/east-africa-drought/east-africa-drought-questions-and-answers. Accessed on 25 Feb 2018

Ataniyazova OA (2003) Health and ecological consequences of the Aral Sea crisis. Prepared for the 3rd world water forum regional cooperation in shared water resources in Central Asia Kyoto, 18 Mar 2003

Economy EC (2013) China's water pollution crisis. Available via https://thediplomat.com/2013/01/forget-air-pollution-chinas-has-a-water-problem/. Accessed on 25 Feb 2018

References

FAO (Food and Agriculture Organisation) (2016) With continued drought, Horn of Africa braces for another hunger season. Available via www.fao.org/news/story/en/item/460996/icode/. Accessed on 25 Feb 2018

FAO (Food and Agriculture Organization) (2017) Lake Chad Basin: a crisis rooted in hunger, poverty and lack of rural development. Available via https://reliefweb.int/report/nigeria/lake-chad-basin-crisis-rooted-hunger-poverty-and-lack-rural-development. Accessed on 25 Feb 2018

Fernandez M (2017) Sri Lanka hit by worst drought in decades. Available via http://www.aljazeera.com/news/2017/01/srilankadrought170122092517958.html. Accessed on 25 Feb 2018

Frederiksen HD (2009) The world water crisis and international security. Middle East Policy 16:76–89

IWMI (International Water Management Institute) (2007) Water matters: news of IWMI research in Sri Lanka, issue 2, March 2007

Jacobs F (2007) The incredible shrinking Lake (Chad, That is). Available via http://bigthink.com/strange-maps/95-the-incredible-shrinking-lake-chad-that-is. Accessed on 25 Feb 2018

Jiang Y (2009) China's water scarcity. J Environ Manag 90:3185–3196

Jing L (2016) Yangtze lakes drying up as China's water crisis spreads. Available via http://www.scmp.com/news/china/policies-politics/article/1903298/yangtze-lakes-drying-chinas-water-crisis-spreads. Accessed on 25 Feb 2018

Mekonnen MM, Hoekstra AY (2016) Four billion people facing severe water scarcity. Sci Adv 2

Ngumbi E (2017) How to tackle repetitive droughts in the Horn of Africa. Available via http://www.aljazeera.com/indepth/opinion/2017/02/tackle-repetitive-droughts-horn-africa-170214090108648.html. Accessed on 25 Feb 2018

Perera A (2016) In parched Sri Lanka, biggest shortage is of water policy: experts. Available via http://www.reuters.com/article/ussrilankadroughtwateridUSKCN12I0PA. Accessed on 25 Feb 2018

Rodrigo C, Senaratne A (2013) Will Sri Lanka run out of water for agriculture or can it be managed? Available via http://www.ips.lk/talkingeconomics/2013/03/22/will-sri-lanka-run-out-of-water-for-agriculture-or-can-it-be-managed/. Accessed on 25 Feb 2018

Shemie D, Vigerstol K (2016) China's has a water crisis—how can it be solved? Available via https://www.weforum.org/agenda/2016/04/china-has-a-water-crisis-how-can-it-be-solved/. Accessed on 25 Feb 2018

Shen Y, Oki T, Utsumi N, Kanae S, Hanasaki N (2008) Projection of future world water resources under SRES scenarios: water withdrawal. Hydrol Sci 53:11–33

Tiezzi S (2014) China's looming water shortage. Available via https://thediplomat.com/2014/11/chinas-looming-water-shortage/. Accessed on 25 Feb 2018

White KD (2013) Nature–society linkages in the Aral Sea region. J Eurasian Stud 4:18–33

WWF (World Wildlife Fund) (2017) Water scarcity. Available via https://www.worldwildlife.org/threats/water-scarcity#causes. Accessed on 25 Feb 2018

Chapter 3
Climate Change: Current Drivers, Observations and Impacts on the Globe's Natural and Human Systems

The earth has always experienced natural climate variability. Scientists, as well as leading scientific organisations, predominantly agree that the climate warming trends observed over the past century are very likely due to human activities, especially in terms of the production of carbon dioxide and other greenhouse gases. Multiple evidence exist which shows that the earth's climate has rapidly changed especially within the past three decades. Evidence collected shows that the warming of the earth is indisputable and needs to be addressed to minimise vulnerability and enable sustainable adaptations. The numerous current climate change drivers, observations and impacts which have been identified first needs to be investigated and understood before one can try to understand how these climate change drivers affects freshwater resources.

Climate change has already been accompanied by various observed impacts on the natural and to some extent human systems. The varied impacts caused by climate change as well as the associated net damage costs will continue to become more significant and increase over time.

3.1 Brief Introduction to Climate Change

The earth's climate has changed throughout the history attributed to natural variability or human activity. The Intergovernmental Panel on Climate Change (IPCC) has defined climate change as the change in climate over time due to natural variability and human activity. This differs from the Framework Convention on Climate Change which attributes the change in climate directly or indirectly to human activities which alter the composition of the global atmosphere, and is an addition to natural climatic variability observed over certain periods. For the purpose of this book, the IPCC definition will be used to define climate change.

In terms of natural climate variability, the earth has experienced 7 cycles of glacial advance and retreat over the last 650,000 years, and an abrupt end to the last ice age approximately 7,000 years ago. This marked the beginning of the modern climate era and human civilisation. Natural climate variability can be attributed to small variations in the earth's orbit which changes the amount of solar energy the planet receives. Currently, the earth is experiencing a warming trend deemed of high significance as most of it is attributed to be human-induced and has proceeded at a rate of unprecedented levels in the past 1,300 years (IPCC 2013).

Scientists, as well as leading scientific organisations, predominantly agree that the climate warming trends of the past century are very likely due to human activities, especially in terms of the production of carbon dioxide and other greenhouse gases. Greenhouse gases (GHGs) are able to affect the transfer of infrared energy through the atmosphere and have heat-trapping capabilities which have led to the significant warming trend. Paleoclimate evidence also shows that the current warming trend is approximately ten times faster than the average rate of ice-age-recovery warming and is of major significance. Further evidence of rapid climate change also includes the following:

- *Global Temperature Rise*: Global surface temperature reconstructions illustrate that the earth has warmed since 1880 with most warming occurring in the past 35 years with 2015 and 2016 being the warmest on record since 2001. Global average temperatures rose by 1 °C or more in 2015 compared to the 1880–1899 average. Even though a decline in solar output has been recorded due to unusually deep solar minimum in 2007–2009, surface temperatures are still on the rise.
- *Increase in Extreme Weather Events*: Extreme weather events such as record high-temperature events as well as intense rainfall events have increased over most of the globe.
- *Decreased Snow Cover*: Spring snow cover in the Northern Hemisphere has shown by satellite observations to be decreasing over the past five decades. The decrease is mainly attributed to snow melting earlier due to increased average temperatures.
- *Glacial Retreat*: Glaciers are retreating in the Alps, Himalayas, Andes, Rockies, Alaska and Africa.
- *Shrinking of Ice Sheets*: Greenland ice sheets have decreased in mass by 150–250 km^3 of ice per annum between 2002 and 2006. The Antarctic has lost approximately 152 km^3 of ice between 2002 and 2005.
- *Declining Arctic Sea Ice*: The extent and thickness of Arctic sea ice have declined rapidly over the past several decades.
- *Warming of Oceans*: Oceans have absorbed most of the heat accompanied with the warming trend.
- *Sea Level Rise*: Global sea level rise of approximately 17 cm in the last century and the last decade has been characterised by nearly double that of the last century.
- *Ocean Acidification*: The acidity of surface ocean waters has increased by approximately 30% since the Industrial Revolution. The increase in acidity is mainly attributed to the increased carbon dioxide emissions into the atmosphere by human activities are being absorbed by oceans and this amount has increased by 2 billion tonne per annum (IPCC 2013).

3.1 Brief Introduction to Climate Change

The scientific evidence which has been collected has therefore shown that the warming of the earth's climate is indisputable and needs to be addressed to minimise vulnerability and enable sustainable adaptations. Before we can discuss the current and projected impacts of climate change on the world's freshwater resources, we first have to obtain a holistic understanding on the current climate change drivers, observations and impacts. This will, in turn, give a detailed understanding of what climate change drivers are affecting freshwater resources, give clarity on the possible interconnectedness of climate change impacts enabling the identification of additional risks and inclusion of applicable vulnerability and adaptation strategies.

This chapter will, therefore, focus on observed changes of climate change over the globe; the main drivers leading to climate change; impacts of climate change on natural and managed resources as well as on human livelihoods with the inclusion of relevant case studies to illustrate its impacts within different contexts. The chapter aims to form the basis of current observations and impacts of climate change around the globe before proceeding to the next chapter which will provide a detailed discussion of the influence of climate change on current and future freshwater resources around the world.

3.2 Observed Changes Within the Climatic System

It has been widely recognised that humans have had a major influence on the earth's climatic system since 1880. Anthropogenic emissions of greenhouse gases are the highest it has ever been and have led to widespread impacts on human and natural systems. IPCC (2014) has stated that it is very likely that the anthropogenic influence, particularly greenhouse gases and stratospheric ozone depletion, has led to detectable observed pattern of tropospheric warming and corresponding cooling of the lower stratosphere.

Evidence indicating the warming of the atmosphere is indisputable. Changes which have been observed since the 1950s have been described as unprecedented over the past decades and millennia. The dominant changes which have been observed in the climate system include the warming of the atmosphere and ocean as well as the diminishing of the amounts of snow and sea level rise (IPCC 2013).

Impacts of climate change have been observed irrespective of its primary cause and have indicated the sensitivity of both natural and human systems to a changing climate. Impacts have been recorded to be more severe and more comprehensive on natural systems and should be monitored more closely. Observed changes can, therefore, be grouped into the following climatic systems, namely the atmosphere, ocean, cryosphere and sea level. This section will provide some background regarding the primary observed changes within the climatic system.

3.2.1 Atmosphere

The primary observation within the atmosphere since 1890 has been the warming of the earth's surface, especially with the last three decades being successively warmer. The warmest 30 year period of the last 800 years in the Northern Hemisphere occurred from 1983 to 2012. The globally averaged combined land and ocean surface temperatures show a warming of 0.85 °C for this period. The total increase between the average of 1980–1900 and 2003–2012 is 0.78 °C. Regional trends for the period of 1901–2012 indicate that the entire globe has experienced an increase in surface temperature which is concerning (IPCC 2013).

The earth's atmosphere is not only characterised by vigorous multi-decadal warming, but globally averaged surface temperatures have also shown substantial decadal and interannual variability. Multiple-independent analyses measurements indicate that the troposphere has warmed and the lower stratosphere has cooled since the twentieth century globally.

3.2.2 Ocean

The ocean is known to be a vital component of the climate system as it regulates the earth's system through redistributing heat around the globe, especially from the tropics to the polar regions, driving climate and weather systems as well as playing a key role in the carbon cycle. It exchanges large quantities of heat, water, gases, particles and momentum with the atmosphere. Oceans play a major role in terms of regulating earth's systems but also supplying living and nonliving resources as well as providing social and economic goods and services.

The primary observed change for oceans has been the overall warming thereof, which in turn dominates the increase in energy stored in the climate system. Between 1971 and 2010, oceans accounted for 90% of the energy accumulated with only 1% stored in the atmosphere. Ocean warming has been recorded to be the largest near the surface with the upper 75 m warmed by 0.11 °C per decade over the period of 1971–2010 (IPCC 2013).

Regions which are characterised by high surface salinity and high evaporation have become more saline. In comparison, regions characterised by low salinity and high precipitation have become fresher since 1950. These trends of ocean salinity provide indirect evidence of changes in terms of evaporation and precipitation over the oceans and consequently for changes in the global water cycle.

The oceanic uptake of CO_2 since the start of the industrial era has resulted in the acidification of the ocean and a decline of ocean surface water pH by 0.1, and a corresponding 26% increase of acidity. Corresponding to warming, a decrease in oxygen concentrations has also been observed in coastal waters and in the open

3.2 Observed Changes Within the Climatic System

ocean thermocline in numerous ocean regions since 1960. Consequently, in recent decades, it is very likely that tropical oxygen minimum zones have also expanded (IPCC 2014).

3.2.3 Cryosphere

The cryosphere, the frozen part of the earth's system or regions where water exists in solid form, is an important part of the global climate system as it acts as a reflective blanket, assisting the earth in not getting too warm. The presence or absence of snow and ice, therefore, affects the heating and cooling of the earth, influencing the planet's entire energy balance. The primary observation over the last two decades has been the loss of mass of the Greenland and Antarctic ice sheets. Glaciers have continued to decrease almost all over the globe and the extent of spring snow cover of the Northern Hemisphere has also continued to decrease. There is high confidence that there are strong regional differences in the trend in Antarctic sea ice extent, with a very likely increase in total extent.

Throughout the twentieth century, glaciers have lost mass and contributed to sea level rise. The rate of ice mass loss from the Greenland ice sheet has very likely substantially increased over the period 1992–2011, resulting in a larger mass loss over 2002–2011 than over 1992–2011. The rate of ice mass loss from the Antarctic ice sheet, mainly from the northern Antarctic Peninsula and the Amundsen Sea sector of West Antarctica, is also likely larger over 2002–2011 (IPCC 2013).

The extent of Arctic sea ice extent has decreased from 1979 to 2012 with a decrease rate in the range of 3.5–4.1% per decade. The extent of Arctic sea ice has therefore decreased in every season and every successive decade since 1979. The most rapid decrease of sea ice sheets occurs in summer where the decrease was in the range of 9.4–13.6% per decade. In terms of the extent of the Antarctic sea ice, it has increased in the range of 1.2–1.8% per decade for the same time period. Strong regional differences do exist with some regions experiencing an increase and others a decrease in the extent of sea ice sheets (IPCC 2013).

Lastly, the snow cover extent of the Northern Hemisphere has decreased since the mid-twentieth century by 1.6% per decade for March and April, and 11.7% per decade for June over the time period of 1967–2012. Permafrost temperatures have increased in most regions of the hemisphere since 1980 with some regions experiencing a reduction in thickness and areal extent. The increase in the permafrost temperatures is mainly due to increased surface temperatures and changing snow cover (IPCC 2013).

3.2.4 Sea Level

The primary observations of sea level have been a global mean sea level rise of 0.19 m over the period of 1901–2010. The rate of sea level rise since the mid-nineteenth century has been larger than the mean rate during the previous two millennia.

Since the early 1970s, glacier mass loss and ocean thermal expansion from warming together explain about 75% of the observed global mean sea level rise. The observed global mean sea level rise has been consistent with the sum of the contributions from ocean thermal expansion due to warming, changes in glaciers, the Greenland and Antarctic ice sheets and land water storage. It should be noted that rates of sea level rise over regions can be several times larger or smaller than the global mean for periods of several decades due to fluctuations in ocean circulation. For example, the regional rates for the Western Pacific are up to three times larger than the global mean, while the Eastern Pacific has experienced rates of near zero to negative. The maximum global mean sea level rise during the last interglacial period has been estimated to be at least 5 m higher than the present, and it did not exceed 10 m above the present. The Greenland ice sheet contributed between 1.4 and 4.3 m to the higher global mean sea level during the last interglacial period, implying an additional contribution from the Antarctic ice sheet (IPCC 2013).

It can therefore be concluded that a wide variety of observations of climate change have been recorded over the last century, and the frequency and magnitude of these observations have been increasing significantly since 1970. The following section will focus upon the primary drivers or attributes of climate change to obtain an understanding of what the main factors are attributing to the observed changes over the globe.

3.3 Primary Drivers

The temperature of the earth is determined by the balance between energy entering and leaving the system. Natural and anthropogenic factors which influence the earth's energy balance includes variations in the sun's energy reaching the planet, changes in reflectivity of the earth's atmosphere and surface and lastly changes in the greenhouse effect which affects the amount of heat retained by the atmosphere.

The earth's climate is influenced by numerous natural factors. Natural drivers of climate change include solar activity, radiative forcings, continental drift, variations in the earth's orbit as well as oceans. These processes have historically altered the face of the planet and occur over very long periods of time. The effects of these natural drivers on the earth's climate have consequently been an alteration between warming and cooling. These natural processes together with the previously mentioned factors as well as other anthropogenic influences such as the cooling effects of aerosol changes, small contributions of ozone changes, land use reflectance changes as well as other minor terms have consequently altered the climate (IPCC 2013).

3.3 Primary Drivers

Both natural and anthropogenic substances and processes are physical drivers of climate change and have changed the earth's energy budget. Radiative forcing is used to quantify the perturbation of energy into the earth's system caused by natural and anthropogenic drivers. Radiative forcings larger than zero lead to a near-surface warming and radiative forcings smaller than zero lead to a cooling. Radiative forcing is estimated based on in situ and remote observations, properties of GHGs and aerosols and calculations using numerical models.

Historical records have established that prior to the Industrial Revolution, the earth's climate was primarily driven by natural factors and processes such as solar energy, volcanic eruptions as well as natural changes in GHG concentrations. Since the Industrial Revolution (1700s), climate changes cannot be explained by natural forces alone as the observed warming, especially since the mid-twentieth century, indicates that human activities have been the dominant cause of warming. The earth's temperatures have increased steadily throughout this period despite natural oppositions such as orbital variations, solar and volcanic activity, El Nino and the Pacific Decadal Oscillation (IPCC 2013).

Anthropogenic GHG emissions have therefore been described as to be the main driver for the warming of the earth's climate as these emissions have increased significantly since the pre-industrial era primarily driven by continued population and economic growth. The highest emissions in history were recorded from 2000 to 2010 and historical emissions have driven atmospheric concentrations of carbon dioxide, methane and nitrous oxide to unprecedented levels of at least the last 800,000 years, and have led to the uptake of energy by the climate system. It should therefore be highlighted that the industrial/anthropogenic contribution of greenhouse gases differs from natural counterparts in major ways and that human influence is occurring much more rapidly, is not cyclical and drives the climate continually into a single direction of the constant uptake of energy by the climate system and consequently warming (IPCC 2013).

This section will, therefore, focus upon providing a description of recent drivers of climate change in terms of natural and anthropogenic radiative forcings, other human activities affecting emission drivers, the attributions of climate change in terms of natural and human influences and provide a brief description of observed impacts.

3.3.1 Natural and Anthropogenic Radiative Forcings

As indicated previously in this section, the atmospheric concentrations of GHGs are at levels which are unprecedented in the last 800,000 years. Carbon dioxide (CO_2), methane (NH_4) and nitrous oxide (N_2O) have all increased since 1750 by 40, 150 and 20%, respectively. CO_2 concentrations are increasing at the fastest observed decadal rate for 2002–2011 and after a decade of stable NH_4 concentrations, measurements have showed renewed increases since 2007, and N_2O concentrations have steadily increased over the last three decades (IPCC 2013).

Total anthropogenic radiative forcing from 1750 to 2011 has had a warming effect and has increased more rapidly since 1970. Carbon dioxide has been the largest single contributor to radiative forcing over 1750–2011 and has been the trend since 1970. The total anthropogenic radiative forcing estimate for 2011 is substantially higher (43%) than the estimate reported in the IPCC Fourth Assessment Report (AR4) for the year 2005. This is mainly due to continued growth in most GHG concentrations and an improved estimate of radiative forcing from aerosols (IPCC 2013).

Radiative forcing from aerosols which includes cloud adjustments has indicated a weaker cooling effect than previously and continues to contribute to the largest uncertainty to the total radiative forcing estimate. Changes in solar irradiance and volcanic aerosol are accompanied with natural radiative forcing. Radiative forcing from stratospheric volcanic aerosols have shown to be associated with a large cooling effect for some years after the volcanic eruption and the changes in total solar irradiance are calculated to have contributed only around 2% of the total radiative forcing in 2011, relative to 1750 (IPCC 2013).

The largest radiative forcing is therefore from well-mixed GHGs which include CO_2, NH_4 and N_2O as well as halocarbons as indicated above. This is followed by other anthropogenic forcings (aerosols, land use surface reflectance and ozone changes), total anthropogenic forcing and natural forcings (solar and volcanic effects).

3.3.2 Human Activities Affecting Emission Drivers

It has been established that approximately half of the cumulative anthropogenic emissions of carbon dioxide between 1750 and 2011, have occurred in the last four decades mostly due to the tripling of emissions from fossil fuel combustion, cement production and flaring as well as the cumulative carbon dioxide emissions which have increased by 40% from forestry and other land use. Approximately, 40% of these anthropogenic emissions have remained in the atmosphere since 1750, and the rest was removed from the atmosphere by sinks and stored in the natural carbon cycle reservoirs. Sinks include ocean uptake and vegetation with soils and account in roughly equal measures. Oceans have absorbed approximately 30% of the emitted anthropogenic carbon dioxide emissions which has caused ocean acidification (IPCC 2013).

Research has shown that the total annual anthropogenic GHG emissions have continued to increase over 1970–2010 with larger absolute increases between 2000 and 2010 indicating that even though mitigating policies have been developed, annual GHG emissions still continued to grow from 2000 to 2010. The global economic crisis of 2007/2008 reduced emissions temporarily and total anthropogenic GHG emissions from 2000 to 2010 were the highest recorded in human history (IPCC 2013).

The main contributors to the significant increase of the total GHG emissions between 1970 and 2010 has been attributed to fossil fuel combustion as well as industrial processes which have contributed 78%. CO_2 remains the major anthropogenic GHG accounting for 76% and fossil fuel-related carbon dioxide emissions have continued to grow. The increase of total annual GHG emissions between 2000 and 2010 is directly attributed to GHG emissions from energy (47%), industry (30%), transport (11%) and building (3%) sectors. GHG emissions have increased in all sectors since 2000 except in agriculture, forestry and other land uses (Shrestha 2016).

Continued economic and population growth are still the main global drivers of increases in carbon dioxide emissions from fossil fuel combustion. The contribution of economic growth increased sharply between 2000 and 2010, while the contribution of population growth remained roughly the same as the previous three decades. Both of these drivers outpaced emission reductions from improvements in energy intensity of Gross Domestic Product (GDP) between 2000 and 2010.

3.4 Attributions of Climate Change

The evidence regarding the human influence on the climate system has continually grown and the influence thereof has been detected in the following:

- Warming of the atmosphere;
- Warming of the ocean;
- Changes in the global water cycle;
- Reductions in snow and ice; and
- Global mean sea level rise.

It is also seen as an extremely likely cause of the observed continued warming since the mid-twentieth century. The changes in the climate in recent decades have been accompanied by numerous impacts on natural as well as human systems worldwide. These impacts have been attributed to observed climate change irrespective of its cause and have highlighted the sensitivity of both natural and human systems to changing the climate.

The attribution of climate change aims to ascertain mechanisms responsible for recent changes in climate. The attributions of climate change to causes enable the quantification of links between observed climate change and human activity as well as other natural climate drivers. The attribution of observed impacts to climate change on the other hand considers the links between observed changes in natural and human systems, and observed climate change regardless of its cause. Therefore, studies which attribute climate change to causes provide information related to the magnitude of warming in response to changes in radiative forcing and enable projections of future climate change while studies attributing impacts to climate change provide indications of the sensitivity of natural and human systems towards future climate. This section will, therefore, focus upon the main natural and influences on climate change as well as the main observed impacts to provide a brief description of attributes of climate change.

3.4.1 Natural and Human Influences

Majority of the research has emphasised the extreme likelihood that the majority of the observed increase in global average surface temperature for the period of 1951–2010 is attributed by the combination of anthropogenic increases of GHG concentrations and other anthropogenic forcings. GHGs have contributed to a global mean surface warming in the range of 0.5–1.3 °C with further contributions from other anthropogenic forcings, including the cooling effect of aerosols from natural forcings and natural internal variability, and have been consistent with the observed warming of 0.6–0.7 °C over this period. GHGs and stratospheric ozone depletion have led to observed tropospheric warming and corresponding cooling of the lower stratosphere since 1961 (IPCC 2014).

Anthropogenic forcings have led to substantial surface temperature increases since mid-twentieth century on every continental region except Antarctica. In terms of Antarctica, large observational uncertainties have resulted in low confidence results which indicate that anthropogenic forcings have contributed to the observed warming averaged over all available stations. However, the substantial warming and resultant loss of sea ice in the Arctic has been attributed to a significant contribution of anthropogenic forcings indicating that human influence has contributed significantly to temperature increases in numerous subcontinental regions as well. Human influences have caused the retreat of glaciers since 1960, have increased the surface melting of the Greenland ice sheet since 1993 and have reduced spring cover in the Northern Hemisphere since 1970 (IPCC 2014).

Research has also established that it is likely that human influences have also affected the global water cycle since 1960 by contributing to increased atmospheric moisture content, to global-scale changes in precipitation patterns over land, to intensification of heavy precipitation over land where there is sufficient data available as well as to changes in surface and subsurface ocean salinity.

Lastly, it has also been established that anthropogenic forcings have contributed substantially to the increases in global upper ocean (0–700 m) heat content since 1970. Evidence of human influences is present in some individual ocean basins. Furthermore, human influences have also been identified as the major contributor to the global mean sea level rise since 1970. This is based on the high confidence in anthropogenic influence on the two largest contributions to sea level rise, thermal expansion and glacier mass loss. The oceanic uptake of anthropogenic carbon dioxide has also resulted in the gradual acidification of ocean sea waters which have led to widespread impacts as well especially for sensitive natural systems (IPCC 2014).

Therefore, it can be concluded that anthropogenic forcings and influences are the primary attribution to the changing of the earth's climate and the continued warming which is accompanied with various and numerous effects or impacts on both natural and human systems. A brief description of observed impacts on these two systems now follows.

3.4.2 Observed Impacts

The changes in the earth's climate have in recent decades been accompanied with impacts on both natural and human systems on all continents and across the oceans. It should be highlighted that the observed impacts of climate change indicate the sensitivity of both natural and human systems towards changing climate irrespective of its cause.

Evidence of climate change impacts is the strongest and most comprehensive for natural systems and some impacts on human systems have been attributed to climate change. The observed impacts on human systems tend to be geographically heterogeneous as they do not only depend on changes in climate variables but also on social and economic factors. Therefore, changes are predominantly observed at local levels while attribution still remains difficult.

The changing of rainfall or the melting of snow and ice has altered the hydrological systems in many regions, and has consequently affected water resources in terms of quantity and quality. Glaciers are continually decreasing in size almost all over the globe due to climate change which has influenced the runoff and water resources downstream (IPCC 2014).

The changing climate has also caused permafrost to warm and thaw in high latitude as well as high elevation regions. The decrease of Arctic sea ice has been one of the observed changes which have been caused by climate change as indicated earlier in this chapter. The Far North region of the world has been warming twice the rate as the rest of the Earth and National Aeronautics and Space Administration (NASA) has been using unique resources to attempt to identify how these drastic changes in Arctic carbon will influence the future climate. Wildfires have been increasing in the North and due to the remoteness of the area causing firefighting to be very difficult, many of these fires burn unchecked for months. It has been estimated that fires have increased by tenfold over the last 50 years in the region and it is still increasing annually. The main concern regarding the increase of wildfires is focussed upon the discharge of huge plumes of carbon into the atmosphere. The permafrost of the Arctic contains more carbon than there is in the atmosphere today (IPCC 2014).

The continued thawing of permafrost will also increase the likelihood of fires but also create atmospheric emissions which will rival fossil fuels through continued decomposition. These emissions will mostly be carbon dioxide and methane and will influence the greenhouse effect. Current studies show that large bursts of methane are already being released from the thawing of Arctic soils.

Biodiversity has also been adversely affected by the changing climate. Numerous terrestrial, freshwater as well as marine species have shifted their geographical ranges, seasonal activities, migration patterns, abundances and species interactions in response to ongoing climatic changes. Even though only a few recent species extinctions have been attributed to climate change, natural global climate change at rates slower than the current anthropogenic climate change have caused significant ecosystem shifts and species extinctions over the past millions of years. Tree mortality has been found to have increased in many places around the world and has been attributed to climate change in some regions. The increases in the frequency or intensity of ecosystem disturbances such as droughts, windstorms, fires and pest outbreaks have been detected in many parts of the world and in some cases are attributed to climate change. Ocean basins have shown changes in abundance, distribution shifts poleward and/or deeper, cooler waters for marine fishes, invertebrates and phytoplankton and alterations have also been observed in ecosystem composition. In terms of some warm-water corals and their reefs, it has been observed that these natural systems have responded to warming with species replacement, bleaching and decreased coral cover causing habitat loss. Most impacts associated with ocean acidification on marine organisms have been attributed to human influence, from the thinning of pteropod and foraminiferan shells to the declining growth rates of corals. Oxygen minimum zones are also progressively expanding in the tropical Pacific, Atlantic and Indian Oceans, as a result of reduced ventilation and oxygen solubility in warmer, more stratified oceans, and are constraining fish habitat (IPCC 2014).

In terms of crop yields, many studies covering a wide range of regions have indicated more negative than positive impacts of climate change. The minority of studies indicating positive impacts relate mostly to high-latitude regions, however, it is still not completely clear whether the balance of impacts has been negative or positive in these regions. Wheat and maize yields have been negatively affected by climate change for many regions and in terms of the global aggregate while the effects on rice and soybean yield have been smaller in major production regions and globally. The observed impacts are mainly focussed upon production aspects of food security rather than access and other components thereof (IPCC 2014).

Lastly, the worldwide burden of human ill health caused by climate change is relatively small compared to the effects of other stressors and is not well quantified. It should be noted that heat-related mortality has increased and cold-related mortality has decreased in some regions due to warming. The local changes in temperature and rainfall have also altered the distribution of certain waterborne illnesses and diseases vectors.

Therefore, climate change has been accompanied by various impacts of varying degrees of magnitude on both natural and human systems. Natural systems have experienced the most widespread effects, however, the effects of climate change on human systems have also been identified with low to medium confidence. Climate change is consequently attributed to both natural as well as human radiative forcings as well as human activities which affect emission drivers. The main culprit for the dramatic trends in warming since 1970 can be attributed to GHG emissions, primarily carbon dioxide, methane and nitrous oxide which are released by a variety of human economic activities. The evidence regarding the human influence on climate change has become undeniable and can be described as an extremely likely cause for the observed warming in the twentieth century. The observed impacts of climate change have varied in terms of frequency and magnitude over the globe and are dependent on context. Impacts of climate change can not be generalised as the characteristics of different regions or areas are not uniform and will be affected in different ways and intensities. The impacts, vulnerabilities and risks in terms of different natural and managed resources now follows to obtain a clearer view of how different elements of the world will be affected by current and continued climate change.

3.5 Impacts, Vulnerabilities and Risks to Natural and Managed Resources

Climate change has already been accompanied with various observed impacts on the natural and to some extent human systems. The magnitude of the effects of climate change also varies according to individual regions and will vary over time. The heterogeneous nature of the effects of climate change over the globe is mainly attributed to the varied societal and environmental systems as well as the varied ability of individual regions to mitigate or adapt to change.

The varied impacts caused by climate change as well as the associated net damage costs will continue to become more significant and increase over time. A good example of this has been the changes in numerous extreme weather and climate events which have been observed since 1950. Some of these events have been linked to human influences which have included a decrease in cold temperature extremes, an increase in warm temperature extremes, an increase in extreme high sea levels as well as an increase in the number of heavy precipitation events in numerous regions around the globe (IPCC 2014).

Recent climate-related extreme events have included heat waves, droughts, floods, tropical cyclones as well as wildfires. The impacts which have been associated with these events have exposed vulnerability as well as exposure of some ecosystems and numerous human systems to current climate variability. Impacts from these extreme climate-related events include the alteration of ecosystems, disruption of food production and water supply, damage to infrastructure as well as settlements, human morbidity and mortality as well as consequences for mental health and human well-being. These impacts have affected all countries of varied levels of development and the magnitude of these effects has been dependent on the preparedness for current climate variability. The impacts of these climate-related extreme events have been severe in some regions and have consequently highlighted the significant lack of preparedness and vulnerabilities within these regions (IPCC 2014).

As indicated previously, the associated net damage costs associated with the impacts of climate change will increase over time. In recent decades, the direct and insured losses from weather-related disasters have increased substantially globally as well as regionally. The major attribute for the increase in net damage costs has mainly been due to the increased exposure of the human population and economic assets to these weather and climate-related disasters.

We, therefore, need to focus upon the impacts on natural and managed resources to determine vulnerabilities as well as risks for improved preparedness to current and future climate variability. For the purpose of this book, focus is placed on the impacts, vulnerabilities and risks on terrestrial and inland water systems as well as food security and food production systems.

3.5.1 Terrestrial and Inland Water Systems

Most ecosystems are vulnerable to changes in climate and react to it in varied ways. Biota and ecosystem processes have been affected by lower rates of climate change which have been projected under high warming scenarios and indicate their high-level vulnerability. The paleoecological record has shown that global climate changes of varying magnitudes under all projected scenarios in the twenty-first century have resulted in large-scale biome shifts and changes in community composition but also possible species extinction in some groups.

Climate change will, therefore, be a major stressor on both terrestrial and freshwater ecosystems in the latter part of the twenty-first century especially in the event of a high warming scenario. The main drivers include direct human activities which mainly includes land use and changes in land use, pollution and water resource development. These human activities are set to continue and dominate the pressures on freshwater resources and terrestrial ecosystems over the next three decades. Continued climate change will also be accompanied by adverse impacts on biodiversity. It should be highlighted that changes resulting from climate change will only be clear after several decades due to the long response times in ecological systems (IPCC 2014).

The substantial alteration of terrestrial ecosystems through changes in plant cover, biomass and phenology or plant group dominance, caused by effects of climate change or through other mechanisms such as land use changes (e.g. conversion of land to agriculture or urban cover) also affects the local, regional and global climates. The feedbacks between terrestrial ecosystems and the global climate, therefore, needs to be considered and includes mechanisms such as changes in surface albedo, evapotranspiration as well as GHG emission and uptake.

Changes in the distribution of freshwater species as well as the degradation of water quality or increase of water quality problems as a result of human activities such as the high anthropogenic nutrient loading will also occur as a result of rising water temperatures due to global warming. Climate change induced changes on rainfall will lead to the alteration of ecologically important attributes of flow regimes in numerous rivers and wetlands but also worsen impacts from human water use especially in developed river basins (IPCC 2014).

Plant and animal species will also be affected through moving their ranges, altering the abundance and shifting seasonal activities in a response to observed changes in climate over the recent decades. General patterns of species and biome shifts towards the poles and higher altitude in response to warming temperatures have been well documented and these general patterns of range shifts have been observed over the last couple of decades in some well-studied specie groups such as insects and birds mainly due to changes in climate. The interactions between the changing temperatures, rainfall and land use can also result in range shifts that are downhill or away from the poles. Other species movement changes include the changes in community composition due to decreases in abundance of some and increases in others, changes in seasonal activity which cause species to change differentially, disrupting life cycles and species interactions. Both of these last mentioned changes will alter ecosystem function and the indicated trends related to plant and animal species will set to continue in a response to future changes in climate.

The slow response times in ecological systems in combination with increased climate velocity, numerous species will be unable to move fast enough during the twenty-first century to track suitable climates under the mid and high-range rates of climate change. Climate velocity which refers to the rate of movement of the climate across the landscape will exceed maximum velocity at which most organisms can disperse or migrate. Populations of species which are unable to keep up with their climate niche will find themselves in unfavourable climate and be unable to get to potentially suitable climate areas. Species located on extensive flat landscapes will be very vulnerable as they will have to disperse over longer distances than mountainous species to keep up with changing climates. Species characterised by low dispersal capacity will also be very vulnerable and include trees, many amphibians and some small mammals. Furthermore, habitat fragmentation, as well as human-made impediments such as dams and urbanised areas, will serve as a further barrier to dispersal as they will reduce the ability of species to migrate to more suitable climates. However, it should be noted that the large intensity of climate change will also reduce populations, vigour and viability of species with spatially restricted populations even if the species have the biological capacity to move fast enough to track suitable climates (IPCC 2014).

The capacity of species to be able to respond to climate change will for many species be constrained by non-climate factors which include but are not limited to the simultaneous presence of inhospitable land uses, habitat fragmentation and loss, competition with alien species, exposure to new pests and pathogens, nitrogen loading as well as tropospheric ozone. In terms of alien species, the establishment, growth, spread and survival of these populations have increased but the ability to attribute this to climate change is low in most cases. These trends are mostly driven by human activities and increased disturbances from natural and anthropogenic events which are in some cases facilitated or promoted by climate change (IPCC 2014).

As stated previously, a large fraction of terrestrial and freshwater species faces increased extinction risk under projected climate change during and beyond the twenty-first century. This is mainly attributed to climate change interacting with other pressures such as habitat modification, exploitation, degradation and invasive species.

Furthermore, increases in the frequency or intensity of ecosystem disturbances such as droughts, windstorms, fires and pest outbreaks have been observed detected in many parts of the world and in some cases can be attributed to climate change. The associated changes in ecosystem disturbances beyond the range of natural variability consequently affects the structure, composition and functioning of ecosystems relatively abruptly and are spatially patchy which in turn affects the distribution and abundance of species.

Lastly, a high risk exists that large magnitudes and high rates of climate change can be associated with low mitigation climate scenarios which will result in abrupt and irrevocable changes in the composition structure and function of terrestrial and freshwater ecosystems on a regional scale. The continuation of climate change will transform species composition, land cover, drainage as well as permafrost extent of the boreal-tundra system which will lead to decreased albedo and the release of GHGs. Adaptation measures will consequently not be able to prevent substantial changes in the boreal-Arctic system. In terms of the Amazon, climate change is not projected to be the leader in abrupt widespread loss of forest cover, but rather a projected increase of severe drought episodes with land use changes and forest fires. The Amazon forest will consequently become less dense, drought- and fire-adapted ecosystems which will put a large number of biodiversity at a higher risk while decreasing net carbon uptake from the atmosphere. Increased reductions of deforestation with wider application of wildfire management are therefore needed to reduce the risk of abrupt changes and accompanied impacts in the Amazon (IPCC 2014).

The implementation of management actions will only reduce and not eliminate the risks to terrestrial and freshwater ecosystems. The capacity for natural adaptation by ecosystems and their constituent organisms is substantial, but for many ecosystems and species, it will be insufficient to cope with projected rates and magnitudes of climate change in the twenty-first century without substantial loss of species and ecosystem services, under medium-range warming. Adoptions within ecosystems can, however, be increased through reducing other stresses, the rate and magnitude of climate change, habitat fragmentation as well as increasing connectivity, maintaining genetic diversity and functional evolutionary processes as well as assisting in the translocation of slow-moving organisms or those whose migration is impeded. It should be noted that in some cases, adaptation responses which have been implemented in agricultural and urban sectors have in some cases led to unintended negative outcomes for terrestrial and freshwater ecosystems. A good example of this has been adaptation responses aimed at countering increased variability of water supply (building more and larger impoundments and increased water extraction) have in some cases worsen the direct effects of climate change in freshwater ecosystems. Therefore, the attempt to mitigate climate change through the widespread transformation of terrestrial ecosystems such as climate sequestration or conversion of previously uncultivated or non-degraded land to bioenergy plantations, will lead to negative impacts on ecosystems and biodiversity. The unintended effects which could accompany adaptation and mitigation measures, therefore, have to be highlighted and kept in mind before it is implemented within terrestrial and freshwater ecosystems to limit further degradation and negative effects of climate change as well anthropogenic processes or activities.

3.5.2 Food Security and Food Production Systems

Numerous definitions exist for food security and have been subject to debate. Earlier definitions focussed upon food production, whereas recent definitions highlight access to food. Food security can, therefore, be defined as that all people have physical and economic access to sufficient safe and nutritious food at all times to meet their dietary needs and food preferences for an active and healthy life. According to the Food and Agriculture Organisation (FAO), estimates showed in 2007 that approximately 75 million people were added to the total number of undernourished. Enough food is, however, produced per capita to feed the global population, yet 870 million people remained in the undernourished category between 2010 and 2012 (FAO 2012). The vast amount of these people lives in developing countries with the largest number (300 million) of food insecure people found in South Asia. Food insecurity is therefore also closely linked with poverty. This has raised the question of how much climate change can be attributed to future changes in the current food production systems and food security.

Negative effects of climate change have been more common than positive ones. The effects of climate change on crop and terrestrial food production has been evident in several regions across the world. Positive trends have been identified in high-latitude regions. Periods of rapid food and cereal price increases which have followed climatic extremes in key producing regions have been observed that highlights the sensitivity of current markets to climate extremes which have been attributed to anthropogenic emissions. Climate trends are also influencing the abundance and distribution of harvested aquatic species, both freshwater and marine, and aquaculture production systems in different parts of the world in a negative manner. This trend is expected to continue with negative impacts on food security, especially for tropical developing countries. However, it may also have positive benefits in other regions which may become more favourable for aquatic food production (IPCC 2014).

The changes in the climate with CO_2 concentrations can also enhance the distribution and increase in the competitiveness of agronomically important and invasive weeds as it will reduce the effectiveness of some herbicides. All aspects of food security potentially be affected by climate change, including food access, utilisation and price stability (IPCC 2014).

Major crops which include wheat, rice and maize planted in tropical and temperate regions will be negatively impacted upon in terms of production by climate change without adaptation in the event of local temperature increases of 2 °C or more above late twentieth-century levels. Individual locations may however benefit. Projected impacts vary across crops and regions and adaptation scenarios, with about 10% of projections for the period 2030–2049 showing yield gains of more than 10% and about 10% of projections showing yield losses of more than 25%, compared to the late twentieth century. The risk of more severe impacts increases after 2050, and show that crop production to be consistently and negatively affected by climate change in the future in low-latitude countries. Climate change may, however, have positive or negative effects in northern latitudes and will also progressively increase the interannual variability of crop yields in many regions (IPCC 2014).

Agronomic adaptations have shown to improve yields by an average of 15–18% of current yields but the effectiveness of these adaptations is highly variable. Benefits of adaptation are higher for crops in temperate rather than tropical regions with wheat- and rice-based food production systems showing larger possibilities of adaptation than maize.

Global temperature increases of approximately 4 °C or more above late twentieth-century levels, combined with increasing food demand would pose significant risks to food security globally and regionally. Low-latitude areas will have greater risks of maintaining food security. Furthermore, changes in temperature and rainfall, without considering effects of CO_2, will also contribute to increased global food prices by 2050 with estimated increases ranging from 3 to 84%. Adaptations implemented in fisheries, aquaculture and livestock production may potentially be strengthened by adoption of multilevel adaptive strategies to minimise negative impacts. Some of the key adaptations for these sectors may include policy and management to maintain ecosystems in a state that is resilient to change, enabling occupational flexibility and development of early warning systems for extreme events. Livestock systems have to implement adaptations cantered around adjusting management to available resources through using better breeds adapted to prevailing climate as well as removing barriers to adaptation. A range of potential adaptation options are available across all food system activities, not just in food production, but benefits from potential innovations in food processing, packaging, transport, storage and trade still have to be looked at and sufficiently researched (IPCC 2014).

The natural and managed resources will, therefore, be impacted upon by climate change in numerous ways all around the world. Impacts vary from low to high intensity or significance as well as frequency. Vulnerabilities to these observed and predicted impacts also vary according to the region or country's characteristics such as population demographics as well as economy and natural resources. Vulnerabilities towards current and future climate change will, therefore, be context and risk dependent and require specific adaption frameworks and strategies to address certain impacts, vulnerabilities and risks. Brief descriptions regarding the impacts as well as vulnerabilities associated with climate change on urban and rural human settlements, human health and livelihoods as well as the economy now follows to establish how the world's human population are and will be affected.

3.6 Impacts and Vulnerabilities of Human Settlements, Health, Livelihoods and the Economy

As described throughout this chapter, climate change will have a wide impact on all natural and managed resources which are driven by natural and human radiative forcings as well as other human activities. The effects of climate change on the world's human population will also be characterised by a variety of impacts which is once again context specific. Brief descriptions of current observed as well as future impacts of climate change on both urban and rural settlements, human health and livelihoods, as well as the economy with the use of some examples and case studies where applicable, now follows.

3.6.1 Urban Settlements

It is estimated that urban settlements house more than half of the world's population and has a primary focus upon its assets and economic activities (World Bank 2008; UN DESA Population Division 2012). Adaptation to climate change, therefore, depends centrally on the actions within urban centres and will require responses on all government levels, individuals, communities as well as the private sector and civil society. The impacts of extreme weather events which have been experienced by numerous urban centres around the world have exposed the risks and vulnerabilities which need to be addressed and have also emphasised that climate change will place additional pressure on these centre's vulnerabilities and risks.

Approximately, more than half of the world's population was living in urban centres in 2008 and this has continued to grow. The world's urban populations and most of its largest cities (75%) are in low and middle-income countries and these nations now hold almost two-fifths of the world's population. This has represented an important and rapid change as previously the highest proportion of the world's urban population was located in highest income nations. The low- and middle-income nations' urban centres include numerous successful cities, however, local governments have not been able to manage the rapid economic and physical expansion of these urban centres which have been accompanied by large deficits in infrastructure and service provision. Additional to this, approximately one in seven people live in poor quality, overcrowded accommodation with inadequate or no basic infrastructure and services, mostly in informal settlements which add health risks and additional vulnerabilities in these settlements. The rapid growth within these highly vulnerable urban communities living in informal settlements has also increased the risk from extreme weather events. These elements as well as the accompanied underlying large and complex economic, social, political and demographic changes have made these urban centres very vulnerable to climate change as these centres are unable to sufficiently adapt to climate change (IPCC 2014).

Risks, vulnerabilities as well as impacts related to urban climate change have been increasing across the globe in urban centres of all sizes, economic conditions and site characteristics. Urban climate change-related risks have been increasing and have included rising sea levels, storm surges and heat stress, extreme rainfall, inland and coastal flooding, landslides, drought and increased aridity, water scarcity as well as air pollution. These risks have been accompanied by numerous widespread impacts on the world's human population in terms of health, livelihoods and assets as well as on local and national economies and ecosystems. People living in informal settlements and hazardous areas which either lack essential basic infrastructure and services or who have the inadequate provision of adaptation have intensified these risks.

The impacts of climate change will affect a broad spectrum of infrastructure systems (water and energy supply, sanitation and drainage, transport and telecommunications), services (health care and emergency services), the built environment as well as ecosystem services which in turn interact with other social, economic and environmental stressors. Consequently, risks are exacerbated and compounded which affects individual and household well-being. City regions and cities also influence the local microclimate as they are sufficiently dense and of a spatial scale. Climate change will also interact with these conditions and may in some cases worsen the level of climate risk. Some of the major impacts related to infrastructure systems, services, the built environment and ecosystem services in terms of urban settlements include the following.

Water Supply, Wastewater and Sanitation: Climate change will impact residential water demand, supply and its management. Projected impacts include altered rainfall and runoff patterns in cities, sea level rise resulting in saline ingress, constraints in water availability and quality, heightened uncertainty in long-term planning and investment in water and wastewater systems. These impacts will interact with growing populations, demand and economic pressures which may increase water stress and negative effects on the natural resource base in terms of water quality and quantity, wastewater and sanitation systems will increasingly be overburdened during extreme rainfall events if attention is not given to maintenance and other management systems.

Energy Supply: Climate change-related impacts on power or fuel supplies will have widespread effects as energy plays a major role in economic development, health as well as quality of life which can have far-reaching consequences on urban businesses, infrastructure, services, residents as well as water treatment and supply, rail-based public transport and road traffic management. Potential impacts of climate change on the energy sector include reductions in efficiency of water cooling of large electricity generating facilities, change in hydro and wind power potential, changing demand for heating or cooling through altering patterns of urban energy consumption, water scarcity or variability can interrupt hydropower supplies and heat waves can result in brownouts or blackouts due to spikes in demand for air conditioning.

Transportation and Telecommunications: Extreme events such as storm surges will affect transportation and telecommunications infrastructure through flooding. Extreme events associated with climate change may also cause disruption of transport as well as a loss of telecommunications which can inhibit disaster response and recovery efforts.

Built Environment including Recreation and Heritage Sites: Extreme events associated with climate change such as tropical cyclones and floods will have adverse effects on urban housing, especially those built with informal building materials and outside safety standards. Increase in wind speeds can also cause widespread damage of structures. Increased climate variability, increased temperatures, altered precipitation patterns and increased humidity will accelerate the deterioration and weathering of stone and metal structures in many cities. Recreational sites such as parks as well as heritage sites such as Venice will also be affected as they may be subject to storm surge flooding.

Ecosystem Services and Green Infrastructure: Ecosystem functions will be affected by climate change through alteration of temperatures and rainfall regimes, evaporation, humidity, etc. It may also highlight the value of ecosystem services and green infrastructure (interventions to preserve the functionality of existing green landscapes and transform the built environment through phytoremediation and water management techniques and by introducing productive landscapes) for adaptation. Influences of climate change may include increased difficulty to expand green infrastructure due to variable rainfall, vegetation such as trees may become more prone to heat stress and pests and urban coastal wetlands may be inundated due to sea level rise.

Health and Social Services: The effects of climate change will be largely felt by the most vulnerable in society which includes children, the elderly as well as the severely disadvantaged. The effects will also be evident across most urban public services which include social care provision, education, police as well as emergency services. Few studies have however been done (IPCC 2014).

Actions within urban centres, therefore, play an essential part in global climate change adaptation as they hold a high proportion of the world's population and economic activities which are most at risk of climate change and also generate a high proportion of global greenhouse gas emissions by urban activities and residents. Urban climate adaptation can consequently build resilience as well as enable sustainable development through delivering mitigation co-benefits as well as resource efficient means to address climate change and sustainable development goals.

3.6.2 Rural Settlements

Approximately 3.3 billion people live in rural areas. Rural areas still account for 47.9% of the world's population where 90% of these live in the developing countries which are described as being less or least developed. The rural population within these nations account for 70% of the world's poor people. Rural areas, especially in developing countries, are dependent on agricultural and natural resources, poverty, isolation as well as lower human development, marginalisation and neglect by policymakers. These characteristics, in turn, make rural areas very vulnerable to the impacts of climate change (IPCC 2014).

Climate change will occur in rural areas in terms of economic, social and land use trends. Observed impacts of climate change include direct impacts such as droughts, storms as well as other extreme events on infrastructure as well as health. Long-term impacts include declining yields of major crops and consequent impacts on livelihoods. Major impacts of climate change will be largely felt on water supply, food security and agricultural incomes. The impacts of climate change will have the largest effect on developing countries due to their economic dependence on agriculture and natural resources, low adaptive capacities as well as geographic locations.

Rural people all over the globe have, however, over long periods of time been able to adapt to climate variability or have learned how to cope with changes in climate. The use of their indigenous knowledge and the development of similar adaptations and coping strategies with supportive policies and institutions can form the basis of adaptation to climate change. However, the effectiveness of these approaches will be determined by the severity and speed of climate change impacts. The development of climate policies which include increasing energy supply from renewable resources, biofuels or payments under reducing emissions from deforestation and forest degradation may have significant secondary impacts which may be positive and negative. Positive impacts may include increase in employment opportunities and negative may include landscape changes as well as the increase of conflicts for scarce resources in certain rural areas. We, therefore, need to be cognisant of how policies may affect rural livelihoods and that secondary impacts and trade-offs may have further implications.

3.6.3 Human Health

The health of the world's human population is very sensitive to changes in weather patterns as well as other changes in climate. The effects on human health can occur both directly and indirectly through changes in climate. Direct effects on human health may include changes in temperature and rainfall as well as the occurrence of heat waves, floods, droughts and fires. Human health may be indirectly affected or damaged by ecological disruptions such as crop failures and changes in patterns of disease vectors or through social responses of climate change which may include the displacement of people following a prolonged drought (IPCC 2014).

Climate change will, therefore, act as an agent which exacerbates already existing health problems. Existing diseases may extend their range into areas which were previously unaffected, however, populations which are already most affected by climate-related diseases will be most vulnerable and at risk. The present worldwide issues of ill health from climate change have been small compared to other stressors. Some effects of climate change may include an increase of risk of heat-related death and illness due to changes in temperature and waterborne illnesses and disease vectors will also change due to altered rainfall patterns. Food production may also be reduced and cause some populations to become vulnerable to food shortages.

Major changes which may occur if climate change continues according to current projections include the following:

- Increased risk of injury, disease and death due to more intense heat waves and fires;
- Greater risk of undernutrition due to diminished food production in poor regions;

- Consequences for health of lost work capacity and reduced labour productivity in vulnerable populations;
- Increased risk of food and waterborne diseases as well as vector-borne diseases;
- Slight reductions in cold-related mortality and morbidity in certain areas due to fewer cold extremes; and
- Geographical changes in food production as well as a decrease in disease carrying vectors as a result of exceedance of thermal thresholds (IPCC 2014).

The impacts on health can be reduced but not totally eliminated in populations which benefit from rapid social and economic development especially in poor and least healthy groups. For vulnerability to be most effectively reduced in the near term, programmes which implement and improve basic public health measures should be the primary focus. An example of the possible effects of climate change on human health can be illustrated by the increase of heatwaves in the Midwest, USA.

Heating Up the Midwest in the USA
An increase of heatwaves has been observed in the Midwest region of the USA and it has been estimated that this will increase in frequency in the coming decades. This is mainly attributed to the combination of general warming as well as intense weather patterns which produce heatwaves. Current projections show that mortality from heat will increase substantially with the accompanied increase in health risks from heatwaves.

Illnesses caused by exposure to high temperatures include heat cramps fainting, heat exhaustion, heatstroke and death (Kilbourne 1997). The Midwest region will become increasingly at risk due to demographic shifts to more vulnerable populations as well as cities' infrastructure not originally designed to withstand the increased severity and intensity of projected heatwaves. Populations with the following characteristics will be facing an increased risk and include the older and younger age, the use of certain drugs which interfere with the body's capability to cope with high temperatures, dehydration, low fitness, excessive exertion, overweight, reduced adjustment to outdoor temperatures, urban populations and lower socio-economic status.

Regions which are currently experiencing or which are already susceptible to heatwaves are projected to experience the greatest increase in heatwave intensity and duration. Regions such as the southern, eastern and southwestern parts of the USA which are already adapted may experience negative effects due to climate change through increased power generation due to increased air conditioning as well as increased vulnerability to heat-related morbidity and mortality. Further adaptations within these regions will be required in terms of infrastructure and public health systems to be able to deal with increased heat stress in the warmer climate.

3.6.4 Human Livelihoods

Before the effects of climate change on human livelihoods can be described it has to be emphasised that livelihoods, poverty, the lives of poor people and inequality all interact with climate change, climate variability as well as extreme events in multi-faceted and cross-scalar ways. The interaction of climate change with the multitude of these non-climatic factors makes the attribution of it very challenging.

It has been established that climate-related hazards may exacerbate other stressors which often have negative outcomes on livelihoods of people especially living in poverty. The subtle shifts and trends to extreme events affect the lives of poor people directly through losses in crop yields, destruction of homes, food insecurity and indirectly through increased food prices. These changes can also lead to shifts in rural livelihoods with mixed outcomes and may cause some people to experience chronic poverty due to being unable to rebuild after extreme events or a series of events. Many of these events which affect poor people are weather related and remain unrecognised in many low-income countries mainly due to short time series and geographically sparse or partial data which inhibit detection and attribution (IPCC 2014).

However, it has been observed that climate change and variability will worsen existing poverty, exacerbate inequalities and will trigger both new vulnerabilities and some opportunities for individuals and communities. Poor communities will not be equally affected as not all vulnerable people are poor. Climate change will, therefore, interact with non-climatic stressors and present structural inequalities to form vulnerabilities.

Climate change will also create new poor between now and 2100 in both developed and developing countries and will jeopardise sustainable development. It will worsen multidimensional poverty in most developing countries and create new poverty in certain countries with increasing inequality in both developed and developing nations. Once again, the implementation of current policy responses for climate change mitigation and adaptation should be carefully considered as it will result in mixed or even detrimental outcomes for poor and marginalised people. An example of the possible effects of climate change on human livelihoods is well illustrated by the vulnerable populations of Africa.

Most Vulnerable are in Africa

The African continent is described as to be one of the most vulnerable continents to climate change and variability which is exacerbated by the interaction of numerous stressors. Major economic sectors are vulnerable to current climate sensitivity and vulnerabilities are aggravated by existing developmental challenges which include endemic poverty, complex governance and institutional dimensions; limited access to capital, including markets, infrastructure and technology; ecosystem degradation; and complex disasters and conflicts which in turn contribute to poor adaptive capacity and increased vulnerability to future climate change.

Even though many African farmers have developed numerous adaptation options for them to cope with current variability, agricultural production and food security in many countries and regions may become compromised due to these options not being sufficient for future changes. The semi-arid conditions in many countries on the continent has already made agriculture challenging and climate change may reduce the length of growing seasons and force some marginal agriculture out of production which will adversely affect food security and human livelihoods in the continent.

Water stress will also impose additional pressures on water availability, accessibility as well as demand on the continent which will be accompanied with numerous human health threats as well as affect the livelihoods of people who are particularly dependent on subsistent farming. Human health has already been compromised by a range of factors in Africa and will be further negatively impacted upon by climate change in the future such as resurgence of malaria in some areas of Southern Africa and the East Africa Highlands (WWF 2006).

Both natural and human systems will be affected and current projections indicate a warming trend particularly in the inland subtropics; frequent occurrence of extreme heat events; increasing aridity; and changes in rainfall—with a particularly pronounced decline in Southern Africa and an increase in East Africa. Rain-fed agricultural systems, on which a large proportion of the region's population depends on, are especially vulnerable and will place agricultural livelihoods in a very precarious position. The consequent increase in the rate of rural–urban migration and the movement of people to informal settlements will in turn expose people to a variety of risks different but not less serious as those they faced in their place of origin which can include outbreaks of infectious diseases, flash flooding of low lying areas as well as increases in food prices. Impacts will, therefore, be felt across all sectors and will likely amplify the effects on human health and livelihoods at varying degrees across the African continent (IPCC 2014).

Climate change may, therefore, have adverse effects on human livelihoods, poverty as well as inequalities around the world. It will exacerbate various stressors which may have negative consequences on human livelihoods especially people living in poverty as a result. Africa will be faced with increased challenges and will affect a large proportion of the region's population which is already vulnerable. Climate change will pose significant risks to human livelihoods especially poor populations which depend on agriculture and who migrate to informal settlements in urban centres.

3.6.5 *Economy*

Numerous key economic sectors are influenced by long-term changes in temperature, rainfall, sea level rise as well as extreme events which are all part of climate change. In most economic sectors, the total impact of climate change may be small relative to the impacts of the other drivers which, for example include population, age, income, technology, relative prices, lifestyle, regulation, governance and many other aspects of socio-economic development which will have an impact on the supply and demand of economic goods and services.

Climate change will affect different economic sectors by varying degrees. Some of the effects of climate change on the main key economic sectors which may include the energy sector, water sector, transportation, tourism and insurance health economic sectors include the following.

Energy Sectors: The effects of climate change will include the reduction of energy demand for heating and an increase for cooling in both residential and commercial sectors. Energy demand will also be influenced by changes in demographics, lifestyles, design and heat insulation properties of housing stock, energy efficiency of heating or cooling devices as well as the abundance of other household electric appliances. Climate change will also affect various energy sources and technologies in different ways depending on the resources, technological processes or locations involved. Gradual changes in climate attributes such as temperature, rainfall, windiness, etc. as well as changes in frequency and intensity of extreme weather events will gradually affect operation over time. Thermal and nuclear power plants will be concerned about the climate-induced changes in the availability and temperature of water for cooling and should, therefore, look at options which will enable them to cope with reduced water availability. This will, however, come at a cost as it will be accompanied with a decrease in efficiency of thermal conversion. Climate change will also affect the integrity or reliability of pipelines and electricity grids as they have been designed and operated for current conditions and design standards may have to be changed for the construction and operation of pipelines and power transmission and distribution lines.

Water Supply Infrastructure and Water Demand: Climate change will have both positive and negative impacts, varying in scale and intensity on water supply infrastructure and water demand, however, the economic implications have not completely been established. Impacts on the economy may include flooding, water scarcity and cross sectoral competition. Flooding can lead to direct impacts in terms of capital destruction or disruption and adaptation through construction and defensive investments. Water scarcity as well as the increased competition for water resources which are driven by institutional, economic and social factors may lead to water not being available in sufficient quantity or quality for some water use sectors or locations.

Transportation Infrastructure: Transport infrastructure may be negatively impacted through the malfunction thereof, if weather is outside of the design range which could happen more frequently due to climate change. All types of transport infrastructure will become more vulnerable due to increased temperature and rainfall extremes as well as changes in freeze-thaw cycles. Transport infrastructure on ice or permafrost will be especially vulnerable.

Tourism Industry: Tourism resorts; especially ski resorts, beach resorts as well as nature resorts will be affected by climate change as tourists spend their holidays at higher or lower latitudes. The demand for outdoor recreation will also be affected due to varying impacts of weather and climate and impacts will vary both geographically and seasonally.

Insurance Systems: The increase of frequency and intensity of weather disasters will lead to an increase in losses and loss variability in various regions. Insurance systems will, therefore, be faced with a challenge to continue to offer affordable coverage while raising more risk-based capital, particularly in low- and middle-income countries.

Health Sector: The increase in frequency, intensity as well as the extent of extreme weather events in combination with the increasing demands of healthcare services and facilities and public health programmes will negatively affect the health sector. Other challenges may also include an increase in disease prevention, infrastructure as well as supplies related to the treatment of infectious diseases and temperature-related events which are associated with climate change (IPCC 2014).

The impacts accompanied by climate change may, therefore, decrease productivity and economic growth but the magnitude thereof has still not been completely quantified. The total global economic impact of climate change is very hard to estimate and furthermore detailed research, data collection and analysis of all key economic sectors such as mining, manufacturing and other services will be needed to assess the overall potential impacts on economic systems and sectors.

References

FAO (Food and Agricultural Organisation) (2012) The state of world fisheries and aquaculture 2012. Food and Agricultural Organization of the United Nations (FAO), Rome, Italy, 209 pp

IPCC (Intergovernmental Panel on Climate Change) (2013) Climate change 2013: the physical science basis

IPCC (Intergovernmental Panel on Climate Change) (2014) Climate change 2014: impacts, adaptations and vulnerability

Kilbourne EM (1997) Heat waves and hot environments. In: Noji EK (ed) The public health consequences of disasters. Oxford University Press, New York, NY

Shrestha HB (2016) Taxation and tax exemption as the policy instruments for emission reduction in the building sector in developed countries. Institute for Technology and Resources Management in the Tropics and Subtropics (ITT)

UN DESA Population Division (2012) World urbanization prospects: the 2011 revision. United Nations Department of Economic and Social Affairs (UN DESA) Population Division, New York, NY, USA

World Bank (2008) World development report 2009: reshaping economic geography. The International Bank for Reconstruction and Development/The World Bank, Washington DC, USA, p 383

WWF (World Wildlife Fund) (2006) Climate change impacts on East Africa: a review of the scientific literature. Gland, Switzerland

Chapter 4
Climate Change and Freshwater Resources: Current Observations, Impacts, Vulnerabilities and Future Risks

Anthropogenic influences have affected the global water cycle since 1960 which has led to various climatic changes over land regions however changes in the hydrological system is not always attributed to anthropogenic climate change. Parts of documented changes are however not solely due to natural variability. Anthropogenic climate change is also one of many stressors of water resources. Observed changes in terms of hydrological changes are mainly attributed to climatic drivers, not all which are necessarily anthropogenic, due to the difficulty and uncertainty regarding attribution.

Numerous observations of varying intensities or significance have been made during the past century and climatic drivers in combination with non-climatic drivers of climate change will continue to have an impact upon the world's hydrological system in varying ways and degrees. Projections show that climate change will decrease renewable water resources in some regions and increase in others with large uncertainties in many of these areas. The implementation of adaptive approaches into water management is encouraged as it can be used to address uncertainties caused by climate change.

4.1 Brief Background

The IPCC (2014) states that it is likely that anthropogenic influences have affected the global water cycle since 1960 which has led to increased atmospheric moisture content, global changes in precipitation patterns over land as well as the intensification of heavy precipitation over land regions. These observed changes in the world's water cycle can lead to diverse impacts as well as risks as they are conditioned by and interact with non-climatic drivers as well as water management responses.

Water acts as an agent who delivers climate change related impacts to the world's human population as well as natural systems and multiple economic sectors. The current and future influence of climate change on the world's freshwater resources is of prime importance as it will have severe consequences on the world's human population, natural systems and economic sectors through making them more vulnerable to water-related hazards and risks and will differ across regions as water is a locally variable resource.

This chapter consequently focuses on the evaluation of current observed changes of the water system, possible drivers or stressors, current and future impacts, vulnerabilities and risks as well as include case studies illustrating the influence of climate change on freshwater resources within different regions or contexts around the world.

4.2 Main Drivers

Observed changes in the hydrological system are not always necessarily due to anthropogenic climate change however parts of the documented changes are not due to natural variability. To determine the attribution to changes in the hydrological cycle, one needs to identify all drivers with confidence levels assigned to each. It is therefore difficult to attribute extreme hydrological events such as floods and droughts solely to climate change however climate change can alter the probability thereof.

Climatic drivers which include precipitation, temperature, sea level, carbon dioxide concentration as well as others which were described in the previous chapter, are all drivers which have an influence on the hydrological cycle. The primary climatic drivers which influence freshwater resources are potential evaporation as well as precipitation (strongly related to atmospheric water vapour content). Temperature has increased with minor changes in surface and tropospheric relative humidity being recorded. There is once again uncertainty regarding climatic drivers due to internal variability of the atmospheric system, inaccurate modelling of atmospheric response to external forcing as well as external forcing itself. It has been estimated that more intense extreme precipitation events are expected mainly due to the projected increase in humidity. Some regions especially the Mediterranean, central Europe, central North America as well as Southern Africa are predicted to have longer and more frequent droughts according to precipitation projections (IPCC 2014).

Anthropogenic climate change is therefore only one of many stressors of water resources. Non-climatic drivers include socio-economic development (population increase, economic growth and development as well as GDP), land use and cover changes (urbanisation, forests, etc.) as well as water demand changes (agricultural, industrial, energy as well as municipal and domestic water use sectors) (IPCC 2014).

Freshwater systems will be strongly impacted upon by demographic, socio-economic as well as technological and lifestyle changes which will change both exposure to hazards and water requirements. Changes in land use are predicted to be the dominant factor to affect freshwater systems in the future. An example of this may be the increase in urbanisation which will cause an increase in flood hazards and a decrease of groundwater recharge. Agricultural land use change is of particular importance especially in terms of irrigation as it accounts for most of the world's water consumption and may severely impact upon freshwater availability for humans as well as ecosystems. Population and economic growth may consequently

4.3 Observed Changes and Attributions

Observed changes in terms of hydrological changes are mainly attributed to climatic drivers, not all which are necessarily anthropogenic, due to the difficulty and uncertainty regarding attribution. Numerous changes have been observed in the twentieth century which has been attributed to various climatic and non-climatic drivers. Observed hydrological changes due to climate change can be grouped into six main categories namely:

- Precipitation, evapotranspiration, soil moisture, permafrost and glaciers;
- Streamflow;
- Groundwater;
- Water quality;
- Soil erosion and sediment load; and
- Extreme hydrological events such as floods and droughts (IPCC 2014).

It has been recorded globally that runoff has changed in the last four decades and have been attributed to climate change and to a lesser extent the increase of CO_2 and land use changes. Reduced runoff has been observed in the Yellow River of China mainly due to increased temperature and where only 35% of the runoff reduction can be attributed to human withdrawals. A decrease in glacier meltwater yield has also been observed and been attributed to glacier shrinkage forced by warming. The decreased dry-season discharge recorded in Peru and as well as the disappearance of the Chacaltaya Glacier in Bolivia have been attributed to decreased glacier extent and ascent of freezing isotherm at 50 m per decade over the last two decades (IPCC 2014).

Anthropogenic greenhouse gas emissions have been attributed to the more intense extremes in rainfall especially over the northern tropic and mid-latitudes. The risk of flooding in England and Wales in 2000 has been attributed to anthropogenic greenhouse radiation and the decrease in the recharge of karst aquifers in Spain during the twentieth century to decreased rainfall and possibly an increase in temperature as well as other confounding factors (IPCC 2014).

Decreased groundwater recharge within Kashmir for the period of 1985–2005 was attributed to a decrease in winter rainfall. Increased temperature as well as rainfall with other confounding factors has been attributed to the increased dissolved organic carbon in the upland lakes of the United Kingdom. An increase of anoxia in reservoirs of Spain especially during El Nino episodes has been ascribed to the decrease in runoff mainly due to a decrease of rainfall and increased evaporative demand. Water quality has also been affected by climate change over the past century. Some of the observations have included the variable faecal pollution in a saltwater wetland in California, USA mainly due to variable storm runoff and precipitation. Nutrient flushing from swamps and reservoirs within North Carolina, USA were attributed to tropical cyclones and lastly increased nutrient content of lakes in Victoria, Australia to increased air and water temperatures (IPCC 2014).

4.4 Impacts and Projected Hydrological Changes

Numerous observations of varying intensities or significance have been made during the past century and climatic drivers in combination with non-climatic drivers of climate change will continue to have an impact upon the world's hydrological system in varying ways and degrees.

The primary trend which has been established regarding freshwater-related risks and climate change is that these will become increasingly significant with the increase in GHG concentrations. Higher emissions of GHGs over the globe will have larger and stronger adverse impacts compared to lower emissions which will be accompanied with less damage and consequently cost less to adapt. For every degree increase of global warming, an estimated 7% of the world's population is projected to be exposed to a decrease of renewable water resources of at least 20% and by end of the twentieth century, 100 years' floods are projected to be three times higher due to very high emissions (IPCC 2014).

Climate change will therefore reduce renewable surface water resources but also groundwater resources in the most dry subtropical region which will consequently increase the competition between water sectors such as agriculture, industry, energy production, settlements and ecosystems. The decrease of renewable water resources as well as the intensification of competition for water resources will affect regional energy, food as well as water security and increase water-related risks for various sectors around the globe. In some cases, water resources are projected to increase especially at high latitudes.

No widespread observations have been made regarding the changes in flood magnitude and frequency due to anthropogenic climate change but projections do show variations in flood frequency. Flood hazards are therefore projected to increase especially in parts of South, Southeast and Northeast Asia, tropical regions of Africa as well as South America. The socio-economic losses from flooding have increased since the mid-twentieth century and this increase is mainly attributed to greater exposure as well as vulnerability across the globe. The risks of floods will consequently increase across the globe partly due to climate change.

Meteorological (less rainfall) and agricultural (less soil moisture) droughts are also estimated to increase in terms of frequency across the globe, more specifically in the dry regions by the end of the twenty-first century. These regions will also experience an increase in the frequency of short hydrological droughts (less surface

4.4 Impacts and Projected Hydrological Changes

water and groundwater). Uncertainties do however exist in terms of the frequency of droughts which last longer than 12 months as these depend on accumulated rainfall over long periods. There is also no conclusive evidence present which indicates that the frequency of surface water and groundwater droughts has changed over the past few decades. The impacts of drought have however increased which has been attributed to the increase of water demand by various water use sectors (IPCC 2014).

Streamflow and water quality will also be negatively affected by climate change and cause the degradation of freshwater ecosystems. Ecological impacts of climate change have been expected to be higher than historical impacts mainly due to anthropogenic alterations of flow regimes through water withdrawals as well as the construction of reservoirs. Raw water quality is also estimated to be reduced due to climate change which will pose risks to drinking water quality even with conventional treatment. This will be mainly caused by increased temperature, increased sediment, nutrient and pollutant loadings from heavy rainfall, reduced dilution of pollutants during periods of droughts as well as the disruption of treatment facilities during floods (IPCC 2014).

Snowfall regions have observed alterations in streamflow seasonality due to climate change. These alterations as well as others will increasingly be observed in future due to climate change. Increasing temperatures have reduced the spring maximum snow depth in recent decades in all snow regions except in very cold regions. The reduction in spring maximum snow depth has caused maximum snowmelt discharge, smaller snowmelt floods, increased winter flows as well as reduced summer low flows. The increase in temperature has also caused river ice in the Arctic to break up sooner than in the past (IPCC 2014).

Glacial regions have also been affected by the increase of temperature. Due to nearly all glaciers being too large for equilibrium in terms of the current climate, a water resource change has occurred during most of the twenty-first century. Changes beyond the committed change are expected with continued warming. Glacial-fed rivers and total meltwater yields from stored glacier ice will also increase in numerous regions in the next decades but decrease thereafter. The continued loss of glacial ice entails a change of peak discharge from summer to spring (excludes monsoonal catchments), and may possibly also reduce summer flows in downstream parts of glacial catchments (IPCC 2014).

In terms of soil erosion and sediment loads, little to no observational evidence is present to indicate that these factors have been altered by climate change. The increase in heavy rainfall as well as temperature has however been projected to change soil erosion and sediment yield in future but the extent of these changes is highly uncertain and will largely depend on rainfall seasonality, land cover as well as soil management practices.

Further projected impacts, vulnerabilities as well as risks in terms of the world's freshwater resources now follow to enable the identification of relevant adaptation or mitigation strategies which can be implemented to minimise further impacts accompanied with climate change.

4.5 Projected Impacts, Vulnerabilities and Risks

Projections of freshwater-related impacts vulnerabilities and risks are largely based on the comparison of historical conditions and are mostly helpful in trying to estimate the human impact on the environment and support adaptation to climate change. Existing climate models show that approximately half of the world's population will be living in high water stress areas by 2030, which includes between 75 and 250 million people in Africa. Additionally, water scarcity will also be exacerbated in some arid and semi-arid regions which will displace between 24 and 700 million people around the world (IPCC 2014).

Current projections estimate that for each degree of global warming, up to 2.7 °C above preindustrial levels, renewable freshwater resources will decrease by at least 20% for an additional 7% of the world's human population. The total number of people who have access to renewable groundwater resources is also projected to decrease quite significantly. The portion of the world's human population which live in river basins will experience an increase of new or aggravated water scarcity which will increase with water scarcity. Projections indicate an increase of 8% at an increase of 2 °C to 13% at 5 °C (Gerten et al. 2013). It is therefore of prime importance that future impacts, vulnerabilities and risks be identified to minimise the effects on the already stressed freshwater resources of the world. The following sections will focus on projected impacts, vulnerabilities as well as risks in terms of water availability as well as water uses which include agriculture, industries or energy production, municipal services or domestic use, freshwater ecosystems and other uses.

4.5.1 Water Availability

Current estimates show that approximately 80% of the world's population is already facing serious threats to its water security according to water availability, water demand and water quality indicators (Vörösmarty et al. 2010). Climate change will exacerbate these threats due to causing the alteration of water availability around the world.

The effects of climate change on future water demand vary and are spread wide across climate and hydrological models. This is mainly due to varied climate models and patterns of projected rainfall changes. Strong consistencies do however exist and these include the following:

- There will be a reduced availability of water resources in the Mediterranean as well as parts of Southern Africa. Much greater variations in terms of projections exist for South and East Asia.
- Runoff may increase in some water-stressed areas and may experience a reduction in the exposure to water resources stress.
- Increase in global mean temperature around 2 °C, above preindustrial and changes in population, will have a greater effect on changes in resource availability than

4.5 Projected Impacts, Vulnerabilities and Risks

climate change. Humans will therefore have a greater effect on water availability than climate change over the next few decades however climate change will regionally exacerbate the effects of population pressures.

- Projections of future water availability are sensitive to climate, population projections, population assumptions as well as the choice of the hydrological impact model which is used to measure stress or scarcity (Schewe et al. 2013).

Projections show that climate change will decrease renewable water resources in some regions and increase in others with large uncertainties in many of these areas. Majority of projections do however show a decrease in many mid-latitude and dry subtropical regions and an increase in high latitude and many humid mid-latitude regions. Even though some projections show an increase in some regions, it should be noted that short-term shortages may occur due to reduced snow and ice storage especially in high latitude areas.

The availability of "clean" water can also be reduced due to the negative impacts of climate change on water quality. The use of groundwater to try and lessen the impacts of water stress should not be considered to be a viable option especially in regions which are projected to experience a decrease in groundwater recharge or renewable groundwater resources. It is projected that more than 10% of the world's population will suffer from a decrease of renewable groundwater resources and it is estimated to rise with an increase in temperature.

4.5.2 Impacts on Water Use Sectors

Agriculture

The agricultural sector is the primary user of water across the world. The water demand and use in terms of food and livestock feed production are governed by crop management and its efficiencies as well as the balance between soil water supply and atmospheric moisture deficit. As a result of these dependencies, changes in climate in terms of precipitation, temperature and radiation will affect the water demand of crops grown through irrigated and rainfed systems.

According to a global vegetation and hydrology model, climate change will barely affect global irrigation demand by 2080 of major crops in regions currently equipped by irrigation systems (Konzmann et al. 2013). There is however a high certainty that irrigation demand will significantly increase in areas which include more than 40% across Europe, USA and some parts of Asia. Other regions which include major irrigated areas such as India, Pakistan and South East China, may experience a slight decrease in irrigation demand mainly due to higher rainfall but only according to some climate change scenarios. Once again it needs to be highlighted that effects on the agricultural water sector will vary largely across regions and will be very heterogeneous.

In terms of soil, it has been estimated that poor soil may not be a limiting factor. The physiological and structural crop responses to CO_2 fertilisation may partly cancel out the adverse effects of climate change through potentially reducing global irrigation water demand (Konzmann et al. 2013). However even though this optimistic trend has been projected, increases in irrigation water demand by more than 20% are still projected in some regions such as Southern Europe. The general trend for future irrigation demand is projected to exceed water availability in numerous areas and should therefore receive attention (Wada et al. 2013).

The increasing variability of rainfall across the globe will especially affect rainfed agriculture and will consequently increase yield variability and differences between rainfed and irrigated land. Water demand of these agricultural water uses should therefore be managed more effectively and inefficiencies should be addressed to limit further water losses especially in terms of irrigated crops. The water demand for rainfed crops should also be reduced through improved management practices. It should be highlighted that unmitigated climate change may however counteract these efforts and places emphasis on addressing future climate change as the continuation thereof at current trends will counteract efforts.

Industries and Energy Production

As indicated in the previous chapter, energy production, especially hydroelectric and thermal power plants as well as the irrigation of bioenergy crops will require large amounts of water. Hydroelectric power generation will be affected by climate change through changing mean annual streamflow, seasonal flows, increase of streamflow variability as well as increased evaporation from reservoirs and changes in sediment fluxes. Impacts of climate change on hydropower plants will largely depend on local changes in these mentioned hydrological characteristics as well as on the type thereof and seasonal energy demand. Regions which are characterised by high electricity demands for heating or for summertime cooling may face challenges due to seasonal streamflow shifts and may require the development and implementation of adaptation of operating rules due to climate change.

In terms of water availability for cooling of thermal power plants, the number of days with a reduced useable capacity is projected to increase in Europe and the USA, owing to increases in stream temperatures and the incidence of low flows.

Municipal Services

Climate change will also cause many challenges for water utilities. The problems with which water utilities may be confronted with include the following:

- The increase in ambient temperatures will reduce snow and ice volumes, increase evaporation rate from water bodies and consequently decrease natural storage of water unless precipitation increases its availability. Water demand will also increase with higher ambient temperatures in conjunction with the competition for this resource.
- Changes in timing of river flows as well as possible more frequent or intense droughts will increase the need for artificial water storage.
- The increase in water temperatures, which encourages algal blooms and increases risks from cyanotoxins and natural organic matter in water resources, will force water treatment facilities to implement additional or new treatment of drinking water.

4.5 Projected Impacts, Vulnerabilities and Risks

- Possible drier conditions will increase pollutant concentrations. This is a major concern as some groundwater sources are already at low quality even when pollution may be natural.
- The increase in storm runoff will increase the loads of pathogens, nutrients as well as suspended sediment.
- The rise in sea level which consequently increases the salinity of coastal aquifers will cause a decrease in groundwater recharge (Bates et al. 2008; Jiménez 2008; van Vliet and Zwolsman 2008; Black and King 2009; Brooks et al. 2009; Whitehead et al. 2009; Bonte and Zwolsman 2010; Hall and Murphy 2010; Mukhopadhyay and Dutta 2010; Qin et al. 2010; Chakraborti et al. 2011; Major et al. 2011; Thorne and Fenner 2011; Christierson et al. 2012).

As indicated previously, climate change will also affect water quality indirectly. Numerous cities rely on water from different forested catchments which do not require a lot of treatment. However, an increase in the frequency and severity of forest wildfires may degrade water quality quite significantly. Numerous drinking water treatment facilities are not designed to handle more extreme effluent variations which are expected to increase under climate change. This will consequently place additional demands on these facilities or even require the development of different infrastructure which is capable of operating for longer which will render the treatment of wastewater to be very costly especially in rural areas. Sanitation technologies vary in terms of their resilience towards climate impacts and the following climatic conditions are of interest. First, wet weather especially heavier rainstorms which cause an increase in amounts of water and wastewater combined in systems for short periods. This is of concern as current designs are based on critical "design storms" of historical rainfall data and will have to be modified. New strategies will also have to be developed in terms of adaptation and the mitigation of urban floods which consider climate change as well as urban design, land use, the heat island effect and topography.

Second, drier weather is also a concern especially in terms of the shrinking of soil which causes water mains and sewers to crack or break, making them very vulnerable to infiltration and exfiltration of water and wastewater. The combination of increased temperatures as well as pollutant concentrations, longer retention times and sedimentation of solids will possibly increase corrosion of sewers, decrease asset lifetimes, increased drinking water pollution and higher maintenance costs.

Lastly, sewer systems are also concerned with a rise in sea levels mainly in terms of the intrusion of brackish or salty water into sewers which necessitates processes that can handle saltier wastewater. The increase in storm runoff also implies that there will be a need to treat additional wastewater when combined sewers are used. Increased storm runoff will therefore add to sewage and the resulting mixture will have a higher pathogen and pollutant content.

Drier conditions will have higher concentrations of pollutants in wastewater and any other type will have to be dealt with. The cost thereof may cause low-income countries to not address this in future. The disposal of wastewater or faecal sludge is a concern that is just beginning to be addressed in the literature and further research is needed for treatment facilities and regions as a whole to be able to adapt accordingly (Seidu et al. 2013).

Freshwater Ecosystems

Freshwater ecosystems include biota (animals, plants, and other organisms) and their abiotic environment in slow-flowing surface waters such as lakes, man-made reservoirs, or wetlands; in fast-flowing surface waters such as rivers and creeks; and in the groundwater. These systems have been heavily affected by human activities than have marine and terrestrial ecosystems.

Climate change will therefore be an additional stressor which will affect these systems not only through increased water temperatures but also through altering stream flow regimes, river water levels as well as the extent and timing of inundation. Wetlands which are located in dry environments are hotspots of biological diversity and productivity, and their biotas will be at great risk of extinction if runoff decreases and the wetland dries up.

These freshwater systems are also affected by water quality changes which may be induced by climate change. Human adaptations to climate change-induced increases of streamflow variability and flood risk, such as the construction of dykes and dams. The effects of climate change on freshwater systems need to receive attention as it supports a wide variety of biota upon which some human livelihoods depend. The degradation of these systems by climate change as well as human activities may have wide-ranging effects on the environment itself as well as the world's human population.

Other Uses

Hydrological changes will also have indirect impacts on navigation, transportation, tourism and urban planning (Pinter et al. 2006; Koetse and Rietveld 2009; Rabassa 2009; Badjeck et al. 2010; Beniston 2012) and may cause social and political problems. An example of this may be where water scarcity and water overexploitation may lead to an increase of risks of violent conflicts and nation-state instability (Barnett and Adger 2007; Burke et al. 2009; Buhaug et al. 2010; Hsiang et al. 2011).

The rise in snowline as well as the shrinkage of glaciers has been estimated to have a very likely environmental, hydrological, geomorphological, heritage and tourism resources impacts in cold regions (Rabassa 2009), as already observed for tourism in the European Alps (Beniston 2012). It should be noted that even though most impacts will have adverse effects, some might be beneficial if managed correctly.

Climate change will therefore be an important influence and constraint of water availability across the globe in the future. We can conclude out of this section that climate change has and will have further negative effects on the hydrological cycle due to continued warming over the recent decades. Changes have included increased atmospheric water vapour content, changes in rainfall patterns including extremes, reduction in snow and ice cover especially in high latitudes and lastly changes in runoff and soil moisture. Substantial changes will therefore be accompanied with future climate change especially under future scenarios of GHG emissions although there are some uncertainties in projected patterns of precipitation on a regional scale.

4.5 Projected Impacts, Vulnerabilities and Risks   65

The relationship between climate and water resources does not exist in isolation and is further influenced by socio-economic and environmental conditions which vary over regions. Multiple human activities especially agriculture, changes in land use, construction as well as water pollution negatively influence water availability. Water demand is also highly variable and is determined by population as well as levels of development. We therefore have to take a closer look on a regional scale in order to determine more specific influences of climate change in certain regions for these regions to be able to develop relevant cost-effective adaptation or mitigation strategies. A brief description regarding the main influences of climate change on freshwater resources in different regions around the world as well as discussion of three case studies to illustrate the wider influence of hydrological changes due to climate change now follows.

4.6 Regional Outlooks and Case Studies

The influence of climate change on the world's freshwater resources will pose significant challenges in terms of adequate future water availability and may pose different risks for socio-economic development as well as environmental sustainability. The impacts of climate change are heterogeneous in nature and will vary according to a region or area's environmental and socio-economic characteristics which in turn determines the degree of vulnerabilities and risks as highlighted throughout. Anthropogenic climate change will pose the biggest challenges to the globe's society and will force decision-making bodies such as governments, to incorporate climate-related risks into their decision-making processes.

This section will focus on providing brief descriptions of the impacts of climate change in terms of regions across the globe and provide detailed case studies for Africa, Pacific Islands as well as Portugal to illustrate various vulnerabilities and risks which may be accompanied with climate change in terms of freshwater resources as well as the different implications thereof on socio-economic as well as environmental systems.

Europe: The European region will experience an increase in the demand for irrigation purposes but also an increase in the frequency of droughts as well as warmer day temperatures. Water made available from runoff will decrease. Some regions are projected to experience a decrease in river flooding however it is projected that the United Kingdom might experience increases. The region is expected to experience an increase in its average crop yields however it should be noted that these projections have a high degree of uncertainty and also assumes that there will be an adequate supply of water for irrigation in the region. Importantly, Europe is an exporter of wheat and maize as well as an importer of all four major crops which will interlink it with climate-related impacts in the Americas as well as Asia in particular which should consequently also be taken into account when looking at future food security within the region.

Middle East and Africa: As indicated previously, the northern regions of Africa as well as the Middle East are already experiencing water stress. The region around the Mediterranean is estimated to experience some of the largest increases in the number of drought days of all regions across the globe. It will experience a decrease in annual water runoff and the warmest days are projected to become even warmer in the already hot climate. The Middle East and Northern regions of Africa are major importers of wheat, maize and rice which will link them to climate-related impacts in major production regions of these crops, namely North America, South America, Russia, Australia as well as Northern Europe.

In terms of the *sub-Saharan region of Africa*, extremely large relative population increases are projected along with decreases in average annual water runoff. This will consequently increase pressure on food and water demand in a region which already suffers from high levels of food insecurity and water stress. The region is also characterised by governance issues where a number of countries have scored high on the Fragile States Index between 2005 and 2013. In terms of future climate in the region, it is expected that the temperatures of warmest days, number of drought days as well as the frequency of flood events are projected to increase across the whole sub-Saharan region which will exacerbate current vulnerabilities such as food insecurity and water stress if adaptation strategies are not developed and implemented.

Asia: For the purpose of describing the effects of climate change in the Asian region, it will be divided into South, East and Southeast Asia. The southern region of Asia is characterised by high population density and the continued population growth will increase demand for food and water resources in a region which is already experiencing water stress and food insecurity. Additional pressure will be placed on the region's already insecure food supply with a predicted decrease in the average yields of wheat and maize. There will be a small increase in rice which is the main export crop for the region. It should be noted that the range spans from 16% decrease to 19% increase in the average yield. The region will also experience an increase in the frequency of flood events in the future as the region is exposed to tropical cyclones along with rising sea levels. The combination of these effects may consequently have adverse effects on the region's population as millions of people may be flooded per year along the coasts.

The eastern region of Asia imports a high proportion of crops which include wheat, maize and soybeans. Over 40% of the world's soybeans are imported by China to try and meet the growing demand for animal feed within the country. These factors consequently link the Eastern Asia region with climate impacts in major production and export regions of these mentioned crops, primarily the Americas. It is also estimated that the region will experience an increase in the frequency of flood events. The region is also exposed to tropical cyclones and characterised by high population near the coast. Rising sea levels will therefore have the potential to affect millions of people as the case in the Southern Asia region. Furthermore, the predicted increase of sea temperatures and ocean acidification may also threaten the region's important fishing industry and place additional pressure on its food supply.

The densely populated south-eastern region of Asia is already exposed to flooding as well as storms. The region is projected to experience considerable increases in population as well as rising sea levels and an increase in the frequency of inland flooding. The increase of sea surface temperatures as well as ocean acidification is also predicted for the region and may threaten fish stocks in this major fishing region. The Southeastern Asia region is also important in terms of rice exports and a major producer of maize. Projections show a slight increase in average rice yield and a decrease in average maize yield. These projections however did not account for increasing water demand for irrigation, decreasing water runoff, increases in drought days as well as the effect of storms.

Australasia: The region which comprises of Australia, New Zealand and New Guinea as well as the neighbouring islands of the Pacific Ocean, is low in population density and has a high level of self-sufficiency in terms of food. The region is however a major exporter of wheat and there are currently mixed and uncertain projections regarding changes in future average yields which depend on adequate water supply for irrigation purposes. The water demand for irrigation has been projected to increase and large increases in the number of drought days and temperature of the warmest days have also been projected. A decrease in water runoff is also projected which may be concerning as it will influence water availability and may place the region's water resources under stress with the increase in water demand especially by the agricultural sector.

North America: This region is very important in terms of crop production as it is the primary source of wheat, maize and soy bean exports to the world market as well as the second largest exporter of rice after Asia. Projections related to future crop yields are currently very uncertain. The region has however shown some increases in the yield of wheat, soybean and rice and decreases in maize yields. Once again, these estimated changes assume sufficient water supply for irrigation as the water demand of the agricultural sector increases. The number of days in drought is also expected to increase as well as the temperature of the warmest days. Projections in terms of flooding are uncertain due to the variability of precipitation estimates.

South America: This region is also very important in terms of crop production, particularly maize and soybeans. Projections show decreases in yield for both these crops as well as wheat in Brazil and northern South America. The more southern regions may have a slight projected increase in these crops. The whole region is however expected to experience reductions in water runoff, increases in the number of drought days as well as higher temperatures, combined with increases in water demand for irrigation (Met Office 2016).

Therefore, the effects of climate change will be heterogenic in nature across regions as indicated above. However, most regions will experience an increase in drought days, decrease in runoff, increase in temperatures of warmest days and an increase of water demand, especially irrigation, as they attempt to achieve future food security. Current vulnerabilities such as water stress as well as food insecurities will exacerbate the effects of climate change, especially in developing regions such as Africa, and may inhibit growth or worsen current socio-economic development and future quality of its natural resources, especially water. The following sections will provide case studies to further illustrate the heterogeneous nature of the impacts of climate change and also highlight the effects thereof on natural resources, specifically water as well as a region or country's socio-economic development and future growth.

4.6.1 Africa: Vulnerable Rivers and Economies

Africa as a region is of a growing concern as its water resources are particularly vulnerable to climate change due to already suffering disproportionately from water-related hazards, primarily floods and droughts (WWAP 2003). The exact magnitude of current water issues on the continent is still not clear however estimations by Vörösmarty et al. (2005) are that approximately 25% of the continent's population experiences water stress and 69% live under conditions of water abundance. These estimations do not consider actual water availability and the relative abundance shows low water consumption which results from limited water supply infrastructure and a further one-third of the continent's population live in drought areas (WWF 2000). Droughts and floods are exacerbated by the relatively low level of economic development and the sub-Saharan region is the only region in the world which has become poorer in the last generation.

The low level of economic growth is clearly shown in terms of the poverty statistics on the continent. The continent makes up 13% of the world's population but holds 28% of the world's poverty, is home to 32 of the 38 indebted poor countries of the world and did not achieve most of the MDGs. The challenge of poverty as well as relatively low economic growth is further intensified through rapid population growth which places further pressure on the continent's natural resources and socio-economic development. The resultant poverty and underdevelopment across Africa has been created by numerous factors such as the difficulty coping with climate variability and changes in terms of frequent droughts, floods, high temperatures, land degradation as well as being dependent on rainfed agriculture which in turn emphasises the need for a clearer understanding regarding the possible changes in the hydrological cycle over Africa related to climate change to inform relevant decision-making. Future general projections of impacts related to climate change for the African continent include the following:

- Warming of more than the global annual mean warming with the drier subtropical regions warming more than the moister tropics.
- Precipitation is likely to decrease over the majority of the Mediterranean region of Africa as well as the northern Sahara. A decrease in winter rainfall over western southern Africa and a likely increase of annual mean rainfall over East Africa.
- Most projections show that annual river runoff will be reduced over North Africa and most of Southern Africa. Increased annual runoff is projected in East Africa.
- The continued increase of the population will result in more water stress in North, Eastern and Southern Africa and the combination of climate change will further increase water stress issues over most of the continent.

4.6 Regional Outlooks and Case Studies

These projected changes and impacts will vary within different regions across the African continent due to varying natural or physical characteristics as well as socio-economic elements such as demographics and level of economic development. Climate change will have a variety of impacts on natural and human systems as described throughout. The projected impacts as well as possible socio-economic consequences for the different regions across the continent also need to be considered to determine what adaptation strategies will be necessary. The impacts as well as their socio-economic effects in terms of the main African regions include the following:

- *West Africa*: Predicted impacts include the risk of rising sea levels and coastal floods to coastlines. These impacts will cause economic damages as well as be associated with rising costs. Changes in terms of disease burden/vector-borne diseases will cause rising health costs. Changes in ecosystem services are also predicted for the region and will negatively affect forestry and fisheries.
- *North Africa*: This region will also be threatened by rising sea levels and coastal floods to coastlines and be accompanied by economic damages and rising costs. Most of the costs will have to be invested in coastal protection as well as identified long-term risks and migration especially in the Nile Delta. The predicted increase in temperature will cause rising summer electricity use as the use of cooling aids increase which will cause higher energy costs. The region is also expected to experience reduced water availability and be accompanied by losses of increased costs of supply. The reduction in agricultural yields and increased irrigation costs will also be accompanied with economic losses. Lastly, the increased health effects of more frequent or intense heat waves will cause an increase in health costs.
- *Central Africa*: The region will experience an increase in health costs due to changes in disease burden or vector-borne diseases. Forestry and fisheries ecosystem services will also be affected.
- *East Africa*: This region will also experience an increased risk of rising sea levels or coastal floods of coastlines such as the Mombasa region. The predicted changes in the frequency and magnitude of extreme events such as floods and droughts will cause economic damages in the region. Changes in disease burden, specifically malaria, may spread to highlands and cause increased health costs. Changes in water availability as well as linkages with lakes and other ecosystems are expected. Predictions also show a potential decrease in agricultural production and there may be a loss of ecosystems as well in the region which may include forests as well as wildlife and tourism which include parks and coral.
- *Southern Africa*: The region will experience increased risk of sea level rise, coastal floods as well as the erosion of coastlines which will cause increased economic damage and costs. The possible decrease in water resource availability will also cause losses and further economic costs for sectors such as agriculture where irrigation is predicted to increase. Changes in disease burden or vector-borne diseases will also cause an increase in health costs. The region is also expected to experience an increase in the risk of forest fires, negative effects on agricultural production as well as the loss of ecosystems or natural resources which could potentially negatively affect wildlife and tourism revenues.

Climate change will therefore have numerous environmental impacts which will consequently lead to various socio-economic costs if proper mitigation or adaptation strategies are not developed. It has been estimated that in the event of no adaptation taking place in the African continent, the total cost may be quite significant. Current estimations regarding the cost to Africa's GDP include a cost range of 1.7% in the event of a 1.5 °C temperature increase, 3.4% in the event of a 2 °C temperature increase and lastly 10% in the event of a 4 °C temperature increase. Cost ranges with adaptation for the region have been estimated to be a minimum of US$10 billion by 2030 and possibly up to US$30 billion a year directly in response to climate change. Africa can therefore not afford to ignore the future effects of climate change and avoid the development of appropriate mitigation and adaptation strategies within different regions and countries (Boko et al. 2007).

Consequently, we need to look on a basin scale to achieve appropriate decision-making especially in terms of smaller basins as global scale analyses only provide a coarse resolution. Basin-scale hydrological models supply more explicit representations of available freshwater resources as well as water demand, providing more detailed evaluation of water availability. Two examples namely the Okavango Delta wetland in Botswana and the Mitano River basin in Uganda will be used to illustrate the importance of basin-scale analysis of climate change and show how a region's characteristics such as the size of the population as well as hydrology influence its vulnerability towards future climate-related changes.

4.6.2 The Future of the Okavango Delta Under Climate Change

The Okavango Delta, one of the most significant hotspots in the Kalahari region of Southern Africa, is an alluvial fan where the river terminates and is one of the major water resources in the region. The Okavango River is one of the largest river systems in Africa and spans over three riparian states of Angola, Namibia and Botswana. The delta is maintained by annual flooding of the Okavango River which consequently creates the second largest inland wetland in the world. The delta is unique and constitutes of a variety of dynamic habitats with exceptionally high beta diversity. It is one of the World Wildlife Fund's top 200 ecoregions, is the world's largest RAMSAR site and became the 1000th UNESCO heritage site in 2014. The delta plays a huge role in the region's hydro-climatology and may therefore be very sensitive to future changes in climate (Murray-Hudson et al. 2006).

The delta supports life in an otherwise inhospitable environment and is home to approximately 150,000 people who live within and around it and depend on the extraction of natural resources either directly or indirectly. Its rich wildlife diversity as well as permanent water resources has consequently been attracting more land users with variety of land use activities. Increased human activities together with the impacts of climate change have caused tremendous environmental stress in the region and raises concerns regarding its future sustainability. Climate change predictions for the Okavango Delta indicate the following:

- Changes in climate even by the middle of the twenty-first century may be potentially very large in the basin and could exceed very substantial natural variability experienced in recent decades;
- Uncertainty exists in terms of the magnitude and signs of climate change but the signal of climate change is clear within the basin; and
- Future variability of precipitation within the basin is still largely uncertain however toward the latter decades of the twenty-first century, there is a convergence of projected responses and shows a drying of the system due to increased temperature and evapotranspiration losses starting to dominate.

The Okavango River basin is one of the least developed river basins in Africa but the increased socio-economic needs of the growing population will change this situation in future and has been identified as a basin which has the potential for water-related disputes (Wolf et al. 2003). Water resources in Angola are particularly exploited and future developments in the country such as increased urbanisation, hydropower schemes as well as increased irrigation will have significant consequences for the water availability in downstream countries especially Botswana and other negative impacts such as the degradation of water quality (Pinheiro et al. 2003; Ellery and McCarthy 1994; Green Cross International 2000; Mbaiwa 2004). Developments in downstream areas must also be kept in mind as the increased water demand will also be an issue. The planned pipeline from the Okavango River to Grootfontein, in an attempt to connect river systems with Windhoek in Namibia, will also exacerbate water demands on the river basin. Developments therefore need to be carefully considered as these will occur within the context of climate variability and change and should be accompanied with the development of appropriate adaptation strategies. The three riparian states which share the water resources of the Okavango River basin also importantly need to have shared accountability regarding future water demands and developments which may affect the future of the river basin's water availability and to ensure its future sustainability under climate variability and change as well as human activities.

4.6.3 *Future of the Mitano Basin, Uganda Under Climate Change*

The Mitano basin in Uganda, in comparison to the Okavango Delta's surface river flow and wetland flooding, is characterised by groundwater recharge. Groundwater is an important water source for drinking and irrigation around the world and supplies 75% of all improved sources of drinking water in sub-Saharan Africa (Foster et al. 2006). The impacts of climate change on groundwater are however poorly understood and most predictions of the effects of climate change on freshwater resources are defined in terms of mean annual river discharge or runoff. An understanding of both is however required for the development of proper climate change adaptation strategies.

This is particularly important for Uganda as the country has a substantial dependence on rainfed agriculture and has a heavy reliance on localised, untreated groundwater as a source of potable water. The Mitano River basin is relatively small and drains areas of relatively high elevation and the groundwater from the weathered overburden and fractured bedrock discharges into the River Mitano drainage network.

The East African region which includes Uganda is one of the few regions in the world which has consistent projection of future precipitation change, with most models showing increased precipitation in the future (Meehl et al. 2007). The main concern is however regarding its future water resources due to projections showing that the population will nearly double between 2005 and 2025. This significant increase in the region's human population will be accompanied with numerous land use changes as well as increased water demands by different sectors which will place immense pressure on the region's surface and groundwater resources. It is therefore of prime importance that climate change related impacts on freshwater resources are projected and considered in an attempt to ensure future sustainability within the region. The following effects of climate change on freshwater resources have been projected for the region:

- An annual increase of 17% is predicted for rainfall except during January and an increase of 4.2 °C in temperature which may give rise to a 53% increase in annual evaporation.
- It is also predicted that these mentioned changes above may lead to a 49% reduction in recharge and 72% increase in runoff.
- Simulations show an increase in the occurrence of large precipitation events (above 10 mm). This increase in intensity under future climate change substantially increases recharge and runoff by overcoming evaporation on individual days and enable more frequent infiltration and recharge.
- Results related to groundwater recharge do however differ between the daily precipitation transformation approach (projected increase) and projections using monthly "change" factors (projected decrease). These results therefore indicate that groundwater recharge is very sensitive to the used projected method.

The projected increase in recharge give a promising outlook for the region's future populations, however the increases in demand due to very rapid population increase will exert significant pressure on its finite water resources. Currently, the country is already experiencing increased motorised groundwater developments which have expanded since 2003 and the main urban area in the Mitano River basin has already been singled out as an area which does not have adequate water supplies to meet the current water demand of the town. Future urban expansions will consequently lead to further intense groundwater abstraction as the country tries to provide safe drinking water to growing urban populations. The uncertainties related to the future development of intensive irrigation under climate change with increased dry-day frequency also poses an additional problem for future water demand in the country.

Socio-economic changes will therefore have a larger effect on the basin's water resources rather than direct climate change impacts. For the country to try and achieve future sustainability of its water resources, it needs to focus on conservative development of groundwater resources especially in terms of intensive groundwater abstraction for town water supplies and irrigation. A full range of possible hydrological responses to future climate change therefore needs to be accounted for when developing future adaptation strategies to be implemented within the Mitano River basin and ultimately Uganda.

4.6.4 Pacific Islands

The Pacific islands is a region constituting of three major island groups in the Pacific Ocean namely Polynesia, Micronesia and Melanesia. The water supplies of islands within the region which are small and low lying will be considerably vulnerable to extreme events such as droughts and may also be threatened by saltwater inundation caused by high tides.

Most of the Pacific islands make use of rainwater catchments as well as shallow wells for drinking water and other uses. On some of the atoll islands, the freshwater layer is thin and very vulnerable to contamination from saltwater below especially in the event of excessive freshwater abstraction. This region has also been affected by El Niño events which have caused severe droughts and water shortages on most of the islands which were accompanied by health effects such as dehydration, drought-related skin disease as well as respiratory infections. Large reverse osmosis water purification systems were consequently invested into address some of the human health concerns. Seawater was therefore treated to assist in alleviating water shortages in conjunction with groundwater withdrawals. Concerns however started to develop during the 1997–1998 drought regarding increased groundwater withdrawals due to the potential impact of saltwater intrusion on different crops such as bananas and breadfruit and groundwater monitoring was implemented. Despite the increase in groundwater withdrawals, saltwater intrusions did not affect crops but emphasised the importance of groundwater monitoring programmes (Keener et al. 2012).

The freshwater resources on the atoll islands are therefore very vulnerable and monitoring data are required to manage both rainwater and groundwater resources conjunctively and increase adaptive capacity of low lying islands to meet challenges from climate variability and change. The predicted temperature increase due to climate change for the region is estimated to be between 2 and 4 °C which will cause immense water stress especially on poor rural people which are dependent on water resources for their livelihoods. The total projected economic loss amount is estimated to be US$1 billion in damages to water resources and will intensify current water problems within the region (Keener et al. 2012).

The major impacts related to projected changes in temperature and rainfall as well as sea level rise will include accelerated coastal erosion, saline intrusions into freshwater lenses as well as increased flooding from the sea which may have large effects on human settlements. Even though it is predicted that rainfall may increase during certain months, the scarcity of freshwater is still a limiting factor in terms of social and economic development. Deteriorating infrastructure is common and results in leakages as well as water pollution and ultimately higher costs.

The population's reliance on rainfall significantly increases their vulnerability in the region to future changes in rainfall distribution. Current estimations indicate that a 10% reduction in 2050 is possible and is expected to correspond to a 20% reduction in the size of the freshwater lens on islands namely Tarawa, Atoll and Kiribati. The declining size of low lying islands due to sea level rise is expected to result in land losses and may also reduce the freshwater lens on atolls by as much as 29% (Keener et al. 2012).

The projected decrease in rainfall as well as accelerated sea level rise would compound together as a great threat to the region's freshwater and groundwater resources as well as its human population. Climate change threatens both natural as well as human systems of the Pacific Islands and will be accompanied with an increase in water stress as well as economic losses in the region. It is therefore of vital importance that the region invests in the implementation of rainwater and groundwater monitoring strategies which may inform the development of appropriate adaptation strategies to ensure future sustainability of the region's water resources but also protect its already vulnerable population against future threats on their livelihoods.

4.6.5 Portugal

Portugal, located in Southern Europe or the Mediterranean region, will also be particularly negatively affected especially in the Iberian Peninsula, located south of the Tagus River. The region is expected to experience a considerable increase in temperature and a reduction in both rainfall and runoff by 2100. Once again, in order to evaluate the impacts of climate change on the region's water resources, one has to focus on changes in runoff and not just temperature and rainfall. Projections show the following for the region:

- A general decrease in the region's water availability;
- An increase of seasonal and spatial asymmetries;
- An increase in flood risk; and
- An increase of water quality problems.

The estimated decrease of runoff in the Spanish part of the transboundary river basins is predicted to accentuate the expected decrease in water availability in Portugal. The influence of climate change on sea levels is also likely to affect groundwater levels as well as quality which will further negatively affect the country's water resource availability (da Cunha et al. 2005).

Some research has already shown small changes regarding the inflow to reservoirs which may cause significant changes in the reliability of water yields from these reservoirs. Changes in operating rules may therefore be required to improve the ability of these systems to meet delivery requirements. The expected changes in rainfall amount and distribution will also have a direct impact on hydropower generation in the country and may also require changes to be made in operating rules. The total impact on the country's agricultural sector will depend on a variety of factors which influence varies for different regions and for different crops. General consensus is however that relatively small changes in water availability could have rather large impacts in the agricultural sector and consequently influence future food security.

Water management in Portugal is therefore likely to become more and more challenging and the impacts of climate change on water demand should be an object of attention as changes in temperature will also have impacts on water demand. The predicted decrease of water availability and increase of hydrological seasonal asymmetries will be associated with increased stressing conditions which relate to water quality and flood risks. Water management policies which are based upon sound and in-depth knowledge of the country's water resources as modified by climate change are therefore of vital importance. Further research related to the impacts of climate change on the region's water resources should be encouraged to enable the consideration of climate change in water management practices. The estimated decrease of river flow in southern Portugal is of primary concern as it may have dramatic consequences on the country's available water resources and emphasises the fact that impacts of climate change on the country's water resources cannot be ignored in terms of planning and management (da Cunha et al. 2005).

Therefore, it is quite clear that climate change will be associated with a wide variety of impacts which need to be considered with an increased attention on the country's water management strategies and policies. Currently, it is said that water managers have not yet fully perceived the current changes and do not assume the need to take climate change into account for long-term planning and the management of their complex water systems. However, some water professional organisations have already perceived the new reality and have consequently made appropriate recommendations to water planners and managers to take climate change into full consideration.

4.7 Conclusions and Future Adaptation

There are numerous opportunities for anticipatory adaptation especially in developing countries and most of the global cost of water sector adaptation will be necessary here. Local level adaptation costs have not been fully established and more research is required.

The implementation of adaptive approaches into water management is encouraged as it can be used to address uncertainties caused by climate change. These adaptive techniques can include scenario planning and experimental approaches which consist of learning from past experiences as well as the development of flexible and low regret solutions which are resilient to uncertainty. Barriers to these approaches do however exist and include the lack of human and institutional capacity, financial resources, awareness and communication.

The increased variability of surface water availability will influence the reliability of water supply which is expected to suffer as a consequence. Increased groundwater abstractions may however assist in continued reliable water supply and may be considered as an adaptation technique but will be limited in regions where renewable groundwater resources are expected to decrease due to climate change.

Certain measures which are aimed at reducing GHG emissions may have risks for freshwater resources as well. Examples of this may include the following. In the event of bioenergy crops being irrigated, these crops make water demands which other mitigation measures might not. Hydropower may also have negative impacts on freshwater ecosystems but can be reduced by appropriate management methods. The capture and storage of carbon can lead to a decrease in groundwater quality and in some regions, afforestation can lead to the reduction of renewable water resources but also flood risks and soil erosion.

Climate change will therefore have varying impacts on freshwater resources as well as at different intensities around the world. The intensity and significance of impacts will largely vary due to different characteristics of regions. Projections show that climate change will decrease renewable water resources in some and increase in others. Uncertainties regarding the actual impact on freshwater resources do however still exist and more research is needed. The development and implementation of various adaptive approaches are however encouraged to try and minimise and address the uncertainties which may be caused by climate change.

References

Badjeck M, Allison EH, Halls AS, Dulvy NK (2010) Impacts of climate variability and change on fishery-based livelihoods. Mar Policy 34:375–383

Barnett J, Adger WN (2007) Climate change, human security and violent conflict. Polit Geogr 26:639–655

Bates BC, Kundzewicz ZW, Wu S, Palutikof JP (eds) (2008) Climate change and water. Intergovernmental Panel on Climate Change (IPCC) Technical Paper VI, IPCC Secretariat, Geneva, Switzerland, 210 pp

Beniston M (2012) Impacts of climatic change on water and associated economic activities in the Swiss Alps. J Hydrol 412:291–296

Black M, King J (2009) The Atlas of water: mapping the World's Most critical resource, 2nd edn. University of California Press, Berkeley, CA, USA, p 128

Boko M, Niang I, Nyong A, Vogel C, Githeko A, Medany M, Osman-Elasha B, Tabo R, Yanda P (2007) Africa. Climate change 2007: impacts, adaptation and Vulnerability. In: Parry ML, Canziani OF, Palutikof JP, van der Linden PJ, Hanson CE (eds) Contribution of working group II to the fourth assessment report of the intergovernmental panel on climate change. Cambridge University Press, Cambridge UK, pp 433–467

Bonte M, Zwolsman JJG (2010) Climate change induced salinisation of artificial lakes in the Netherlands and consequences for drinking water production. Water Res 44:4411–4424

Brooks JP, Adeli A, Read JJ, McLaughlin MR (2009) Rainfall simulation in greenhouse microcosms to assess bacterial-associated runoff from land-applied poultry litter. J Environ Qual 38:218–229

References

Buhaug H, Gleditsch NP, Theisen OM (2010) Implications of climate change for armed conflict. In: Mearns R, Norton A (eds) Social dimensions of climate change: equity and vulnerability in a warming world. The World Bank, Washington DC, USA, pp 75–102

Burke MB, Miguel E, Satyanath S, Dykema JA, Lobell DB (2009) Warming increases the risk of civil war in Africa. Proc Natl Acad Sci USA 106:20670–20674

Chakraborti D, Das B, Murrill MT (2011) Examining India's groundwater quality management. Environ Sci Technol 45:27–33

Christierson BV, Vidal J, Wade SD (2012) Using UKCP09 probabilistic climate information for UK water resource planning. J Hydrol 424:48–67

da Cunha LV, de Oliveira RP, Nascimento J, Ribeiro L (2005) Impacts of climate change on water resources: a case study on Portugal. In: Presented at the fourth inter-celtic colloquium on hydrology and management of water resources, Portugal

Ellery WN, McCarthy TS (1994) Principles for the sustainable utilisation of the Okavango Delta ecosystem: Botswana. Biol Conserv 70:159–168

Foster S, Tuinhof A, Garduno H (2006) Groundwater development in Sub-Saharan Africa. A strategic overview of Key issues and major needs. Sustainable groundwater management, lessons from practice, Case profile collection number 15

Gerten D, Lucht W, Ostberg S, Heinke J, Kowarsch M, Kreft H, Kundzewicz ZW, Rastgooy J, Warren R, Schellnhuber HJ (2013) Asynchronous exposure to global warming: freshwater resources and terrestrial ecosystems. Environ Res Lett 8:034032

Green Cross International (2000) The Okavango river basin. Transboundary basin sub projects: the Okavango. Available via www.gci.ch/GreenCrossPrograms/waterres/pdf/WFP_Okavango.pdf. Accessed on 26 Feb 2018

Hall J, Murphy C (2010) Vulnerability analysis of future public water supply under changing climate conditions: a study of the Moy Catchment, western Ireland. Water Resour Manage 24:3527–3545

Hsiang SM, Meng KC, Cane MA (2011) Civil conflicts are associated with the global climate. Nature 476:438–441

IPCC (Intergovernmental Panel on Climate Change) (2014) Climate change 2014: impacts, adaptations and vulnerability

Jiménez BEC (2008) Helminths ova control in wastewater and sludge for agricultural reuse. In: Grabow W (ed) Water and health Vol. II, encyclopedia of life support systems (EOLSS). Developed under the auspices of UNESCO, Eolss Publishers, Oxford, UK, pp 429–449

Keener VW, Marra JJ, Funicane ML, Spooner D, Smith MH (2012) Climate change and Pacific Islands: indicators and impacts. Pacific Islands Regional Climate Assessment

Koetse MJ, Rietveld P (2009) The impact of climate change and weather on transport: an overview of empirical findings. Transp Res Part D: Transp Environ 14:205–221

Konzmann M, Gerten D, Heinke J (2013) Climate impacts on global irrigation requirements under 19 GCMs, simulated with a vegetation and hydrology model. Hydrol Sci J 58:1–18

Major DC, Omojola A, Dettinger M, Hanson RT, Sanchez-Rodriguez R (2011) Climate change, water, and wastewater in cities. In: Rosenzweig C, Solecki WD, Hammer SA, Mehrotra S (eds) Climate change and cities: first assessment report of the urban climate change research network. Cambridge University Press, Cambridge, UK, pp 113–143

Mbaiwa JE (2004) Causes and possible solutions to water resources conflicts in the Okavango river basin: the case of Angola, Namibia and Botswana. Phys Chem Earth 29:319–1326

Meehl GA, Stocker TF, Collins WD, Friedlingstein P, Gaye AT, Gregory JM, Kitoh A, Knutti R, Murphy JM, Noda A, Raper SCB, Watterson IG, Weaver AJ, Zhao Z-C (2007) Global climate projections. In: Solomon S, Qin D, Manning M, Chen Z, Marquis M, Averyt KB, Tignor M, Miller HL (eds) Climate change 2007: the physical science basis, contribution of working group I to the fourth assessment report of the intergovernmental panel on climate change. Cambridge University Press, Cambridge and New York, NY

Met Office (2016) Regional case studies. Available via https://www.metoffice.gov.uk/climate-guide/climate-change/impacts/human-dynamics/case-studies. Accessed on 26 Feb 2018

Mukhopadhyay B, Dutta A (2010) A stream water availability model of upper Indus basin based on a topologic model and global climatic datasets. Water Resour Manage 24:4403–4443

Murray-Hudson M, Wolski P, Ringrose S (2006) Scenarios of the impact of local and upstream changes of climate and water use on hydro-ecology in the Okavango Delta. J Hydrol 331:73–84

Pinheiro I, Gabaake G, Heyns P (2003) Cooperation in the Okavango river basin: the OKACOM perspective. In: Turton A, Ashton P, Cloete E (eds) Transboundary rivers, sovereignty and development: hydropolitical drivers in the Okavango river basin. African Water Issues Research Unit/Green Cross International/University of Pretoria, Pretoria, pp 105–118

Pinter N, Ickes BS, Wlosinski JH, van der Ploeg RR (2006) Trends in flood stages: contrasting results from the Mississippi and Rhine River systems. J Hydrol 331:554–566

Qin B, Zhu G, Gao G, Zhang Y, Li W, Paerl HW, Carmichael WW (2010) A drinking water crisis in Lake Taihu, China: linkage to climatic variability and lake management. Environ Manage 45:105–112

Rabassa J (2009) Impact of global climate change on glaciers and permafrost of South America, with emphasis on Patagonia, Tierra del Fuego, and the Antarctic Peninsula. Dev Earth Surf Process 13:415–438

Schewe J, Heinke J, Gerten D, Haddeland I, Arnell NW, Clark DB, Dankers R, Eisner S, Fekete B, Colón-González FJ, Gosling SN, Kim H, Liu X, Masaki Y, Portmann FT, Satoh Y, Stacke T, Tang Q, Wada Y, Wisser D, Albrecht T, Frieler K, Piontek F, Warszawski L, Kabat P (2013) Multi-model assessment of water scarcity under climate change. In: Proceedings of the national academy of sciences of the United States of America

Seidu R, Stenström TA, Owe L (2013) A comparative cohort study of the effect of rainfall and temperature on diarrhoeal disease in faecal sludge and nonfaecal sludge applying communities, Northern Ghana. J Water Clim Change 4:90–102

Thorne O, Fenner RA (2011) The impact of climate change on reservoir water quality and water treatment plant operations: a UK case study. Water Environ J 25:74–87

van Vliet MTH, Zwolsman JJG (2008) Impact of summer droughts on the water quality of the Meuse river. J Hydrol 353:1–17

Vörösmarty CJ, Douglas EM, Green PA, Revenga C (2005) Geospatial indicators of emerging water stress: an application to Africa. Ambio 34:230–236

Vörösmarty CJ, McIntyre PB, Gessner MO, Dudgeon D, Prusevich A, Green P, Glidden S, Bunn SE, Sullivan CA, Liermann CR, Davies PM (2010) Global threats to human water security and river biodiversity. Nature 467:555–561

Wada YD, Wisser S, Eisner M, Flörke D, Gerten I, Haddeland N, Hanasaki Y, Masaki FT, Portmann T, Stacke Z, Tessler Z, Schewe J (2013) Multi-model projections and uncertainties of irrigation water demand under climate change. Geophys Res Lett 40:4626–4632

Whitehead PG, Wilby RL, Battarbee RW, Kernan M, Wade AJ (2009) A review of the potential impacts of climate change on surface water quality. Hydrol Sci J 54:101–123

Wolf AT, Yoffe SB, Giordano M (2003) International waters: indicators for identifying basins at risk. In: UNESCO-IHP-VI, Technical documents in hydrology, PC-CP, Series, no 20, p 30

WWAP (World Water Assessment Program) (2003) Water for people, water for life. UN World Water Development Report, WWAP, UNESCO, Paris

WWF (World Water Forum) (2000) The Africa water vision for 2025: equitable and sustainable use of water for socio-economic development. World Water Forum, The Hague, p 30

Chapter 5
Primary Water Quality Challenges, Contaminants and the World's Dirtiest Places

Water has been placed on the forefront of some global goals, numerous agendas and been recognised as vital resource internationally. However it is still being widely degraded at different intensities and rates and remains a significant global challenge in terms of availability but also quality.

Poor water quality directly translates into decreasing the availability of the world's water resources and attributes to water scarcity. Water scarcity caused by increased poor water quality has been exacerbated by increased pollution mainly caused by the disposal of large quantities of insufficiently treated or untreated wastewater. Emerging pollutants in combination with future changes in climate patterns will also exacerbate current and create new water quality challenges, which still have unknown long-term impacts on the world's ecosystems and human health. The world's water resources are being degraded by numerous factors. Surface and groundwater resources as well as the environment and human population as a whole are becoming more and more threatened by numerous types of contaminants as well as toxic pollution problems which may have immediate as well as significant long-term effects.

5.1 Brief Background

Water has been placed at the forefront of global discussion since the Earth Summit in 1992 where it was highlighted as a topic which should receive worldwide attention. These discussions consequently highlighted freshwater's key role in terms of global development, health as well as sustainability which led to the calling for action in several international summits. Freshwater, in general, has, therefore, been globally recognised as a vital resource and strong messages have emerged from summits; however the global challenge of poor water quality has not received as much attention. The development of the MDGs in 2000 primarily focussed upon the urgent need for improvements to drinking water and sanitation but did not specifically emphasise the need for improvements in water quality. Even though adequate quantities of good quality freshwater are critical for human health, food security as well as the aquatic environment itself, the need for improved water quality was still not placed at the forefront of global goals (UNEP 2016).

This has fortunately changed 20 years after the Agenda 21 with the development of the outcome document 'The Future We Want' at the United Nations Conference on Sustainable Development. The Rio +20 summit placed water, in general, at the forefront of sustainable development but recognised it as being linked to numerous critical global challenges, and for the first time highlighted the need for the reduction of water pollution, the improvement of water quality and the reduction of water loss.

Goal 6 of the 17 adopted SDGs focusses on ensuring the availability and sustainable management of water and sanitation for all and specifically calls for sustainable withdrawals, access to adequate quantity for all as well as an improvement of water quality by 2030. The need for the protection of aquatic ecosystems as well as the preservation of ambient water quality was, therefore, properly recognised for the first time as a global issue which needs widespread attention.

This chapter will consequently focus upon water quality as a global challenge by looking at the main causes and contaminants which contribute to the degradation of the world's water quality as well as indicate which regions or areas suffer from immense water quality problems or pollution.

5.2 Water Quality Challenges

Suitable or good water quality together with adequate water supply is necessary to achieve the set SDGs for health, food and water security across the globe. Water quality is one of the biggest global challenges and has been recognised as such.

Societies will increasingly face water quality challenges during the twenty-first century which may threaten human health, limit food production, reduce ecosystem functions and hinder economic growth which will consequently directly translate into environmental, social and economic problems. Poor water quality also directly translates into decreasing the availability of the world's water resources and attributes to water scarcity as a whole. Water scarcity caused by increased poor water quality has been exacerbated mainly through increased pollution of freshwater resources mainly caused by the disposal of large quantities of insufficiently treated, or untreated, wastewater into rivers, lakes, aquifers and coastal waters. Emerging pollutants which include personal care products and pharmaceuticals, pesticides, and industrial and household chemicals, in combination with future changes in climate patterns may also exacerbate current and create new water quality challenges which still have unknown long-term impacts on the world's ecosystems and human health (UNEP 2016).

The world is, therefore, faced with current water quality challenges which have numerous different impacts on ecosystems as well as human health but might also face new challenges in future. This section will, therefore, first focus on the main cul-

5.2 Water Quality Challenges   81

prits attributing to the current pollution of the world's freshwater resources, followed by a discussion related to the main contaminants which contribute to the world's current water challenges and which have developed the world's dirtiest places and rivers. A description of the main causes of water pollution around the world now follows.

5.2.1 Primary Causes of Pollution and Main Water Quality Problems

Freshwater systems are influenced by both natural and human factors around the world. Natural sources of pollution are mainly natural processes and animals which introduce pollution into the water which does not come from human activities. In terms of natural influences, the climatological and geochemical location of the water body gets influenced through temperature, rainfall, leaching as well as runoff of elements from the Earth's crust.

Natural sources of pollution which include both processes and animals are mostly of a non-point pollution source. Examples of natural non-point pollution include the following:

- Some trees and marshes can naturally produce organic matter and lower dissolved oxygen when plant material and parts fall off into waterbodies or get washed into a waterbody through runoff or stormwater.
- Nutrients (nitrogen and phosphorus), which are required by both animals and plants to grow, can have a large impact on the natural balance of freshwater systems as they cause plankton to grow excessively. Plankton also dies excessively and consequently puts a large amount of organic matter into the waterbody resulting in lower dissolved oxygen. Nutrients are naturally recycled from plant to animal, plankton to fish; however excessive nutrients in the natural system cause excessive growth and imbalances. Animals which live in large numbers in waterbodies, also cause direct pollution through their manure which is accompanied by excessive nutrients and possible eutrophication.
- Sediment can also contribute to pollution through sedimentation. Under natural conditions, sediment in water is usually related to big storm events such as tropical storms. It should be noted that it is difficult to determine whether it is caused naturally or by human activities and one has to consider land use patterns.
- Animals can also cause microbial pollution of water bodies. Animals, such as ducks and geese, which live on water in large numbers put manure directly into the water and cause pollution which consequently contaminates the water with disease-causing organisms.

Most surface water bodies are however affected to some extent by impacts from human activities which may include the discharge of waste products or the addition of sediments, salts and minerals with runoff from agricultural areas and urban settlements. The main culprits of the degradation of water bodies correspond with the main water uses and consequently include agriculture and other land-based activities, industrial activities and associated emerging pollutants as well as domestic and municipal wastewater.

5.2.1.1 Agricultural Sector and Other Land-Based Activities

The continued growth in the world's human population has been accompanied with the necessary escalation in agricultural production in an attempt to meet the growing demand for food. This has consequently also been accompanied with an increased pressure on water resources as irrigation increases as well as the degradation of surface water quality through runoff which carry fertilisers and pesticides from agricultural land. Groundwater is also contaminated through the infiltration of agricultural chemicals. The abstraction of water for irrigation purposes may also contribute to the reduction of available surface water volumes and may cause water pollution by returning water that is contaminated by salts, fertilisers and pesticides. The increase in salts as well as minerals will damage the ecosystem and make the water less or unsuitable for certain uses such as irrigation.

The increase in nutrients which arise from land-based activities results in eutrophication and other related changes to aquatic systems in terms of both freshwater and coastal areas. The levels of nutrients, specifically nitrogen and phosphorous, are usually low as it naturally arises from the leaching of minerals or from the decomposition of organic matter. Human activities, particularly, fertiliser runoff from agricultural activities as well as discharge of sewage effluents typically cause higher levels which enhances productivity. The resulting eutrophication leads to the enhancement of productivity within the water body which may have negative consequences for water use in terms of growth of nuisance plants and algal blooms, alterations in the ecosystem structure and fish species as well as the deoxygenation in the event of the decomposition of algal blooms. Rivers are also affected and may carry pollutants which include sediments and nutrients to the coastal zone and the eutrophication of coastal waters. Eutrophication was already highlighted in 1988 to be a global problem and the increasing demand for food production as well as wastewater disposal over the decades have kept it as a major global water quality issue.

Wastewater from various land-based sources such as agriculture, industry and domestic sewage also acts as a pollutant by carrying dissolved salts as well as minerals to surrounding surface waters. The concentrations of these salts will vary according to the source but high concentrations may disrupt the natural chemical balance of the receiving waterbody which leads to salinization and makes the water unfit for certain uses such as industrial use but also unsuitable for aquatic organisms.

5.2.1.2 Industrial Sector with Its Accompanied Emerging Pollutants

The industrial sector which is strongly linked with social and economic development requires access to adequate quantities of suitable quality water for manufacturing

5.2 Water Quality Challenges

and production processes but also for the treatment and assimilation of wastewater products. This requirement is well illustrated in the production of pharmaceutical compounds which requires very high-quality waters as well as in the case of paper production where a large quantity of water is required.

If the current trends in water use and degradation continue, it will be a major challenge to meet the future needs of industries. It has been noted and highlighted numerous times since 1970 that the discharge of waste products, especially those containing metals and chemical compounds, has an impact on the aquatic environment especially in terms of the aquatic food chain as well as predatory animals and birds. Metals which include (but not limited to) mercury, arsenic and cadmium as well as most synthetic organic compounds such as pesticides, are toxic at high enough concentrations to living organisms which includes humans.

Currently, new forms of toxic compounds or emerging contaminants are being evaluated in terms of their potential to reach and affect aquatic environments or human health through the consumption of contaminated water. Some of these new emerging compounds may include excreted and metabolised pharmaceutical products such as hormones and different classes of drugs.

5.2.1.3 Domestic and Municipal Sector Wastewaters

Domestic and municipal wastewaters primarily cause water quality degradation through human sewage which contains pathogens, organic matter as well as chemicals used by people which may include pharmaceutical products. Both human and animal excreta may contain a variety of transmissible pathogens such as viruses, bacteria, protozoa and worms, and the occurrence of these are dependent on the geographical location as well as the occurrence of the disease in the local population.

The organic matter contained in sewage or manure and food waste is naturally degraded within water bodies through chemical and biological activity. However, these processes make use of the dissolved oxygen in water, consequently causing stress for numerous aquatic organisms which include fish species that require oxygen for respiration. The oxygen demand within the water body is therefore used as an indicator of organic matter pollution through the Biological Oxygen Demand (BOD) test which is used to determine the impact of sewage released into rivers. Organic pollution may lead to anoxia or deoxygenation in which few organisms can survive and the consequent low dissolved oxygen may lead to chemical reactions which may result in the release or formation of other toxic compounds which include ammonia and hydrogen sulphide in sediments and bottom waters.

The primary human activities affecting the world's resources are therefore agriculture, industry and energy production, mining activities, water-system infrastructure, uncontrolled disposal of human waste or sewage, the combination of population growth, urbanisation and development and lastly climate change which has been discussed at length in Chap. 4. An overview of the world's primary water quality problems as well as of main contaminants in the water now follows to obtain a clearer view of what some of the current threats and also an overview of the main contaminants in water which lead to widespread water degradation around the globe.

5.2.2 Primary Water Quality Challenges and Contaminants

The quality of the world's water resources has been heavily influenced by human activities, especially in the last three decades. In terms of global water pollution, the primary pollution problems are related to sewage, industrial and agricultural waste as well as wastewaters. It is estimated that 2 million tonnes of sewage, industrial and agricultural waste are discharged every day into the world's water resources. In terms of wastewater, over 80% is not collected or treated worldwide and urban settlements are the main source. The United Nations (UN) estimates that approximately 1,500 km^3 of wastewater is produced annually which is six times more water than exists in all the rivers of the world (UN WWAP 2003). Both aquatic and marine ecosystems are under threat from various human-made pollutants which include chemical spills, industrial waste discharge, acidified rain, sewage overflows as well as fracking fluids. It is estimated that the world's industries release 300–400 million tonnes of heavy metals, solvents, toxic sludge as well as other wastes into water resources annually.

It is of further concern that the pollution in the majority of rivers in Latin America, Africa and Asia have worsened since 1990 and actions will, therefore, need to be taken to try and achieve the SDGs and the Post 2015 Development Agenda. The main water quality problems on these three continents include the following:

- *Severe pathogen pollution*: One-third of all rivers are affected. Hundreds of millions of people are at risk of negative health effects by coming into contact with these rivers where the most vulnerable are women and children.
- *Severe organic pollution*: One-seventh of rivers are affected. The state of freshwater fishery is of major concern as it is important for food security and livelihoods especially for countries that are reliant on inland fisheries. The primary population group which is affected is local poor rural people relying on freshwater fish as their main source of protein in their diet as well as on inland fishery for their livelihood.
- *Severe to moderate salinity pollution*: One-tenth of rivers are affected. High salinity levels may impair the use of water for irrigation, industry and other uses.
- *Increased anthropogenic loads of phosphorus*: Majority of the major lakes affected. May accelerate eutrophication and disrupt natural processes (UNEP 2010).

The increased water pollution experienced in these three continents is mainly attributed to the increase of wastewater loadings to rivers. The main drivers for the increase in water pollution can be attributed to population growth, increased economic activities, intensification and expansion of agriculture as well as the increased untreated sewage. China is also facing major water quality challenges. Approximately 40% of China's rivers have been found to be seriously polluted and 20% too toxic for human contact.

Therefore, above and beyond the pollution of industries and agriculture, untreated sewage is also a major source of continued water pollution. Approximately 80% of untreated sewage is discharged directly into communal waterways in some developing countries. Water quality is, therefore, a major issue in developing nations where it is estimated that 3.2 million children die each year due to unsafe drinking water and inadequate sanitation. These countries are characterised by very limited access to wastewater treatment facilities which leads to waterbodies being used as open sewers for human waste products as well as garbage. The Ganges River in India is a prime example of this where the river receives over 1.3 billion L of domestic waste, 260 million L of industrial waste, runoff from 6 million tonnes of fertiliser and 9000 tonnes of pesticides as well as thousands of animal carcasses (Dakkak 2016).

The primary reason for the lack of adequate water treatment facilities and regulations within developing nations can be attributed to the lack of available finances to fund the development of proper infrastructure which can regulate these pollution sources. Degraded water costs countries in North Africa and the Middle East between 0.5 and 2.5% of the GDP per year further limiting the available funds which can be used to finance necessary wastewater treatment facilities.

The degradation of water resources places immense pressure on the world's available water resources in terms of supply. Water becomes unsuitable for various uses or activities. It, therefore, reduces the amount of clean water which is available for human consumption, sanitation, agriculture and industries as well as to various other ecosystem services. The decrease in the amount of water available for use also holds devastating environmental, health and economic consequences which may put various sectors at risk and disrupt a country's social and economic growth. We, therefore, have to be aware of the main contaminants of water resources in order to be able to address the degradation of the world's water resources and limit further water quality problems.

Water quality is primarily affected by changes in nutrients, sedimentation, temperature, pH, heavy metals, non-metallic toxins, persistent organics and pesticides as well as organic factors. These changes cause alterations in the physical, chemical and biological characteristics of a waterbody and may have specific implications for human and ecosystem health. These contaminants can combine synergistically which result in worse or different impacts than the cumulative impacts of a single contaminant. The continued pollution or input of contaminants results in the resilience of an ecosystem being exceeded leading to non-linear changes which may be irreversible. The current primary contaminants affecting most of the world's water resources include the following.

Nutrients
Nutrient pollution is one of the most widespread environmental problems across the globe and also one of the most challenging and costly problems to address. The problem is primarily caused by excessive nitrogen and phosphorous in water as well as air. The primary sources for nutrient pollution include, but are not limited to, the runoff

of fertilisers, animal manure, sewage treatment plant discharges, stormwater runoff, failing septic tanks as well as car and power plant emissions. These sources consequently cause the nutrient enrichment of water bodies which leads to the increase of primary productivity rates to excessive levels, leading to the overgrowth of vascular plants, algal blooms as well as a decrease in dissolved oxygen causing stress for aquatic organisms.

Algal blooms are visible as bright green and foul-smelling sheen on the water's surface and when these blooms eventually die off, the decomposition consumes dissolved oxygen and leads to low oxygen concentrations in the water column. The decrease in dissolved oxygen to unacceptable levels may kill invertebrates and fish and may develop dead zones where oxygen is so low that no or very little life can be supported. Prime examples of dead zones include the Gulf of Mexico which is caused by agricultural runoff from the Mississippi River watershed (EPA 2012; Beaudry 2017).

Certain algae, like cyanobacteria, may produce toxins which may affect humans and animals who ingest or are exposed to waters characterised by high levels of algal production. In some cases, water treatment can lead to unintended negative effects and conditions when chlorine interacts with algae and produces carcinogenic compounds. Nutrient pollution is, therefore, a primary threat for ecological as well as human health but also the economy in terms of financial losses.

Erosion and Sedimentation

Erosion as a natural process provides sediments and organic matter to water systems, however, human activities have altered natural erosion rates as well as the volume, rate and timing of sediment entering waterbodies and rivers. Anthropogenic erosion and sedimentation has been confirmed as a global issue which is primarily associated with agriculture in the broadest context as the sector is mostly responsible for most of the global sediment supply to water bodies, rivers and ultimately into the oceans. Forestry practices as well as machinery which damage low-growing vegetation can also lead to erosion as these activities leave soil unprotected.

Sediment pollution is characterised by two dimensions namely physical and chemical. The physical dimension is where the loss of topsoil and land degradation caused mainly by gullying and sheet erosion can lead to excessive levels of turbidity in receiving waters. This may in turn cause off-site ecological and physical impacts through deposition in these affected waters leading to high levels of sedimentation and physical disruptions of hydraulic characteristics. The chemical dimension is where silt and clay factions are the primary carrier of absorbed chemicals which include phosphorous, chlorinated pesticides and most metals. These chemicals are transported by sediment into aquatic systems and may lead to negative effects such as decreasing or impairing the spawning habitats of fish, plants and bottom-feeding invertebrates and altering the water chemistry. The turbidity caused by fine particles cause water to become less transparent and blocks sunlight. Decreased light in the water column, may impede the growth of aquatic plants which in turn may affect the habitat of numerous aquatic species such as young fish. Deposited sediment may also smother the gravel beds where fish lay eggs, may bury some aquatic invertebrates who are in a sessile state and also damage their fragile filtering systems. Sediment is ultimately transported to coastal zones and consequently also affects marine invertebrates, fish as well as coral (Castro and Reckendorf 1995; UNEP 2010).

Erosion is mainly very costly to the agricultural sector as the loss of topsoil, in turn, represents an economic loss through the loss of productive land but also by the loss of nutrients and organic matter which needs to be replaced by fertiliser at a considerable cost to maintain soil productivity. It is estimated by the Environmental Protection Agency that sediment pollution causes a total of US$16 billion of environmental damage annually and it's the most common pollutant in rivers, streams lakes and reservoirs. It is further estimated that natural erosion produces 30% of sediment within the USA and accelerated erosion from human use of land for the remaining 30% (Castro and Reckendorf 1995; UNEP 2010).

Erosion and sedimentation of waterbodies and waterways, therefore, have widespread threats and effects on both ecological and socio-economic spheres in terms of negative environmental effects as well as economic losses especially in the agricultural sector.

Water Temperature

Thermal pollution is not the most thought of type of water pollution. The temperature of a waterbody is, however, one of the most important characteristics of an aquatic system as it can affect numerous aspects. Thermal pollution can be defined as the sudden increase or decrease of temperature of a natural waterbody by human influence.

Alterations in temperature may affect an aquatic system through the following:

- *Decreases Dissolved Oxygen*: An increase of temperature may cause a decrease in solubility of oxygen.
- *Increases the Rate of Chemical Processes*: An increase in temperature causes the rate of chemical reactions to increase. It affects the solubility and reaction rates of chemicals in the waterbody.
- *Disrupts/Alters Biological Processes*: The metabolism, growth and reproduction of aquatic organisms may be influenced negatively.
- *Affects Species Composition*: Some species can only survive within a limited temperature range.
- *Influences Water Density and Stratification*: Differences or changes in water temperature and densities between the layers of the water column may lead to stratification and seasonal turnover.
- *Trigger Environmental Cues for Life-History Changes*: Changes in water temperature may act as a signal for certain aquatic insects to emerge or for fish to spawn (RAMP 2008).

The main sources of thermal pollution include power, manufacturing and industrial plants which use water as a cooling agent and consequently releases water back to the source with a higher temperature. Soil erosion causes water bodies to rise and become more exposed to sunlight and consequently lead to higher temperatures which may be fatal for aquatic biomes and lead to anaerobic conditions.

The removal of trees and plants consequently exposes water bodies to direct sunlight, causing increased temperatures as more heat is now being absorbed. Urban runoff discharged into surrounding water bodies can also cause an increase in water temperature especially during summer months where paved areas get hot. Thermal pollution can also be caused naturally through volcanoes and geothermal activities under the ocean and seas which trigger the rise of lava. Lighting can also play a role by introducing huge amount of heat especially into oceans.

Thermal pollution of water bodies leads to a decrease in dissolved oxygen levels, an increase of toxins, loss of biodiversity, alterations of aquatic species' reproductive systems, metabolic rates and migration patterns and lastly has a significant ecological impact through sudden thermal shocks which may cause mass killings of fish, insects, plants or amphibians. Organisms can also become stressed when temperature is too cold or hot, which decreases their resistance to pollutants, diseases as well as parasites. Thermal pollution will, therefore, have a significant negative impact on the health of the aquatic system and changes the overall quality of the affected waterbody. Thermal pollution mainly has environmental impacts and do not have impacts on a region's socio-economic development and growth (UNEP 2010).

Acidification

Acidification of waterbodies takes place over a period of time. The pH of waterbodies is predominantly determined by surrounding soil and rock types and it also determines the health and biological characteristics of aquatic ecosystems. The natural buffering capacity of a waterbody usually neutralises the additional acidity which enters it but when this buffering capacity runs out the acidity can increase rapidly.

Acidification is chemically characterised by the loss of acid-neutralising capacity which in turn decreases the pH and increases the concentrations of sulphate, nitrate and ammonium as well as aqueous aluminium, but also manganese, zinc and other metals within the affected waterbody. These chemicals and metals increase in forms which are available and can accumulate in biota while being directly toxic (Campbell and Stokes 1985) for example the discovery of mercury in acid-stressed aquatic food chains. Acidification can be permanent or chronic in nature, mainly associated with long-term acidification in severely impacted areas or be episodic, short term and reversible in the event of rapid discharges during snowmelt and heavy rainfall.

The main sources of acidity come predominantly from a wide range of industrial activities such a mining and power production from fossil fuels on a localised level. Acid rain can also contribute to acidification and can affect large regions by falling on poorly buffered substrates or geologically sensitive areas such as parts of Scandinavia, central and Eastern Europe, the United Kingdom as well as Precambrian Canada and the USA (Wright 1983).

The best-known sufferers of acidification are fish populations. Fish populations have been found to be physiologically affected by acidification. Most losses of fish populations have been attributed to reproductive failure rather than mortality in adults mainly due to the younger life stages being more sensitive. It should be noted that the mortality of adults also occurs especially in the event of acid-shock episodes. As indicated previously, acidification can cause an increase in metals within an affected waterbody. Large increases of toxic metals such as aluminium ions have been found to reduce the survival and growth of larvae, older stages of fish as well as lead to mortality of the fish population regardless of any direct effect of H^+ (UNEP 2010; Zielinski 2015).

Acidification has immense impacts and consequences on individual fauna and flora as well as for the human population and ecosystems as a whole. It affects all aquatic organisms negatively from diatoms to fish. Affected waterbodies or streams are characterised by having extremely low pH and may appear clear, however closer inspections such as looking under stones will reveal impoverished invertebrate colonisation.

Salinity
Salinisation is referred to as the increase in the concentration of dissolved solids in waterbodies, detected primarily with an increase in chloride which is an important anion of salts. It can also be called the amount of total dissolved salts which are usually but not limited to sodium and chloride ions as well as others such as potassium and bicarbonate ions (UNEP 2010).

The main cause for the salinization of waterbodies has always been related to agricultural activities especially in arid and semi-arid environments throughout the world. The long-term salinization of surface waters due to the increase of roadway coverage as well as urban development have shown to cause a sharp increase in the concentration of sodium as well as chloride in aquatic systems and should also be considered as a dominant source.

Therefore, the main sources of salinization of water bodies can be fertiliser, storm water runoffs as well as urban wastewater discharge. These activities lead to an increase in the concentrations of sodium, sulphate, chloride as well as magnesium. Various other actions, often but not exclusively human activities, also cause the accumulation of salts in waterbodies. These include agricultural drainage from high-salt soils, groundwater discharge from oil and gas drilling as well as other pumping operations, industrial activities and in some cases municipal water treatment operations. Droughts can also exacerbate the accumulation of salts which in turn may also affect the spectrum of industries which depend on clean water (Canedo-Arguelles et al. 2016).

Increased salinity also increases the stress of freshwater organisms by affecting metabolic function and oxygen saturation levels. Riparian and emergent vegetation can also be altered by rising salinity. Characteristics of natural wetlands and marshes can also be altered, the habitat of some aquatic species can be reduced and agricultural productivity and crop yields can also experience a decrease (Carr and Neary 2008).

Increased salinity has been found to cause adverse effects on human health and ecosystem functioning. Water with chloride concentrations above 250 mg/L are associated with a salty taste and may be accompanied with higher sodium concentrations and possibly toxic impurities which is a human health concern. Increased salinity, additionally, can also have immense economic costs as a result of the loss of ecosystem services as well as direct costs related to water treatment for human consumption. Salinisation, therefore, affects both freshwater fauna and flora as these species often do not tolerate high salinity levels, affects human health and ecosystems negatively and may have adverse economic costs or impacts as a whole.

Pathogenic Organisms

Bacterium in water or pathogenic organisms includes bacteria, protozoa as well as viruses which are some of the most widespread and serious type of water contaminants. Major pathogens include viruses, bacteria, parasitic protozoa as well as helminthes. Pathogenic organisms are a major public health hazard with risks factors in most parts of the world as it occurs in most types of water sources. Areas with limited access to safe and clean water are especially affected and are rampant in areas characterised by large amounts of untreated wastewater.

Animal husbandry operations like grazing can also result in high concentrations of bacteria and nitrates and can result in health hazards due to the presence of pathogens. Nitrate concentrations above 10 mg/L can result in unsuitable drinking water and lead to methemoglobinemia (blue baby/) in infants (Hubbard et al. 2004).

Other wastes discharged from agricultural, industrial and domestic activities into surrounding surface waters can lead to other significant environmental and human health concerns (Adams and Kolo 2006). Pathogens of concern which are transmitted through faeces mainly cause gastrointestinal symptoms such as diarrhoea, vomiting and stomach cramps. Others may also cause symptoms involving other organs and severe sequels or be an interrelated factor for malnutrition.

Water that looks clear can also contain pathogens due to raw waste and pollutants being discharged into relatively pristine waterways through non-point and point pollutants. Groundwater wells, lakes as well as water parks and public pools are also at risk of *Escherichia coli* and *Cryptosporidium* pathogens.

Heavy and Trace Metals as well as Other Inorganic Contaminants

The increased contamination of the environment by metals has in recent years grown in concern due to its increasing effects on ecological health as well as global human health concern. Human exposure to these effects has increased dramatically due to the exponential increase of their use in agricultural, industrial, domestic and technological applications around the world.

Aquatic environments are therefore mostly polluted by inorganic chemicals which include heavy metals as well as inorganic anions and radioactive materials. Some trace metals such as arsenic, zinc, copper and selenium are found naturally in different waters around the world. Metals such as cobalt, copper, chromium, iron, magnesium, manganese, nickel, selenium and zinc are all essential nutrients which are required for different biochemical and physiological functions of plants and animals and the inadequacy of these micronutrients may result in a variety of deficiency diseases and/syndromes (WHO 1996). Heavy metals are also considered as trace elements due to their presence in trace concentrations in multiple environmental matrices. The bioavailability of these heavy metals are influenced by physical factors (temperature, phase association, adsorption and sequestration), chemical factors which influence speciation as well as biological factors (species characteristics, trophic interactions, and biochemical/physiological adaptation) (Verkleji 1993; Hamelink et al. 1994; Kabata-Pendia 2001).

5.2 Water Quality Challenges

Anthropogenic activities such as mining, industry and agriculture play a role in the mobilisation of trace metals out of soils and waste waters into surrounding fresh surface waters. The main sources of heavy metal and trace metal environmental pollution include geogenic, industrial, agricultural, pharmaceutical and domestic effluents as well as atmospheric sources. These are predominantly point source pollutants which mostly originate from areas such as mining, foundries and smelters as well as other metal-based industrial operations. Metal corrosion, atmospheric deposition, soil erosion of metal ions and leaching of heavy metals, sediment resuspension and metal evaporation from water resources to soil and groundwater are also considered to be sources of environmental contamination. Natural sources of heavy metal pollution also exist and may include phenomena such as weathering and volcanic eruptions (Tchounwou et al. 2012). Examples of industrial sources may include but not limited to metal processing at refineries, coal burning in power plants, petrol combustion, nuclear power stations, plastics, textiles, microelectronics as well as wood preservation and paper processing plants. Industrial pollution is mainly caused by waste which contains but not limited to mercury, cadmium, chromium and cyanide and domestic pollution through waste which contains mostly nitrogen compounds as well as copper, iron, lead and zinc through the distribution of drinking water. The main inorganic pollutants found around the world as well as their primary reason for concern include the following:

- *Arsenic*: Occurs mainly in groundwater from both natural sources and human activities. It can cause skin lesions, circulatory problems, nervous system disorders and prolonged exposure may cause various forms of cancer.
- *Barium*: Occurs naturally at low concentrations in groundwater. May cause nervous and circulatory system problems such as high blood pressure.
- *Copper*: Originates from corrosion of copper plumbing systems. Levels above 1.3 mg/L are of concern as it may cause severe stomach cramps and intestinal illnesses.
- *Iron*: Is a common known natural problem in groundwater which may be exacerbated by mining activities. It is not considered as a health concern to humans but high concentrations can cause drinking water to have a metallic taste and orange-brown stains.
- *Lead*: It's a primary health concern as it poses serious health threats to safe drinking water. Long-term exposure to lead concentrations above the acceptable standard has been connected to numerous health effects such as cancer, stroke as well as high blood pressure. It can also cause premature birth, seizures, brain damage and lowered IQ in children (Tchounwou et al. 2012).

Heavy metals have been considered to be a group of contaminants which threaten both aquatic organisms and humans for a long period of time even at trace concentrations. These pollutants are toxic and persistent in the environment as they have the tendency to bio-accumulate in living tissues such as fish. Accumulated concentrations can compromise normal physiological processes, impair reproductive and other function and introduces an introductory pathway into the human food chain.

The use of metal polluted water for irrigation purposes may also lead to the death of crops or interfere with the absorption of essential nutrients.

The high degree of toxicity, especially of arsenic, cadmium, chromium, lead, and mercury, place these systemic toxicants as a top public health concern even at lower levels of exposure. Metal polluted water is therefore toxic to aquatic organisms as well as humans even at extremely low concentrations and is a significant health risk and/or concern which needs to be monitored and addressed.

Introduced Species and Other Biological Disruptions

Introduced species or invasive alien species have begun to receive more attention due to them posing a greater risk for biodiversity, human health as well as economies. Ecosystems are degraded and weakened by pollution, climate change as well as fragmentation. The introduction of alien species has placed more pressure on these already vulnerable ecosystems which is very difficult to reverse. A rising incidence has been recorded of invasive species displacing endemic species, altering water chemistry and affecting local food webs. Invasive species are therefore characterised by biological disruptions which in turn affects freshwater systems and should consequently be considered as a water quality problem (UNEP 2010).

In most cases, aquatic species have been introduced deliberately into certain distant ecosystems for specific purposes such as recreational, economic and other purposes. On many occasions, these introductions have devastated endemic fish species as well as other aquatic organisms leading to additional biological disruptions and the degradation of the local or affected watershed. Species which have invaded unintentionally, have been transported on the hulls of recreational watercrafts or in the bilge-water of commercial boat traffic.

Aquaculture, a type of agriculture dealing with the rearing of fish, has been identified as a factor which exerts a variety of impacts on the surrounding environment. Some of these impacts include the large-scale introductions of fish species into areas outside their native range which has led to the emergence of feral populations. Aquaculture can also degrade the host environment, disrupt the host community, creates competition with existing species as well as cause predation and possible elimination of the local biodiversity. Risks associated with the introduction of exotic fish through aquaculture include genetic degradation of the host stock, stunting, the deterioration of indigenous stock, introduction of diseases and parasites and ultimately have socio-economic consequences (UNEP 2010).

The introduction of Water Chestnuts into Lake Champlain can be used as a good example to illustrate both environmental and socio-economic impacts which may be associated with the introduction of alien invasive species into an aquatic environment. This invasive species has had negative impacts in terms of aesthetic values in the case of Lake Champlain. The introduction of water chestnuts (*Trapa natans*) led to negative effects on the Lake's ecosystems and aesthetics by forming a dense mat and root system covering large areas of the surface water and consequently hinders sunlight. These mats have been problematic for recreational activities such as boating, fishing, water skiing which may all hurt property values. It has been estimated that the introduction of water chestnuts have caused the decline of over US$12,000 for shoreline property values especially in heavily infested areas (Eyres 2009; Domske and O'Neil 2003). In terms of environmental impacts, the Water Chestnut mats also deplete oxygen within the waterbody which may cause fish kills. These mats also trap organic matter creating silt and consequently increases sediment levels and breeding spots for mosquitoes. It also increases the growth of algae as well as duckweed.

It should be highlighted that intensive aquaculture carries a great risk of serious aquatic disease outbreaks and causes a greater need for water treatment chemicals as well as drugs for diseases and treatment. Consequently, this could, in turn, lead to the development of resistant strains of human pathogens in adjacent waters which should be considered as a major concern. One of the most dangerous effects of invasive alien species on humans may be their characteristic of being a disease carrier which may lead to resistant strains of human pathogens which may cost affected regions huge amounts if not acknowledged and addressed properly.

Persistent Organic Pollutants and Other Toxins
Persistent Organic Pollutants (POPs), can be described as organic compounds which resist environmental breakdown via biological, chemical and photolytic processes and some may take up to a century to degrade. These persistent human-produced organic chemicals enter both surface and groundwater through human activities such as industrial processes and pesticide use as well as through breakdown products of other chemicals. POPs can be transported long ranges via wind and ocean currents to regions where they have never been produced and are commonly found in groundwater through leaching through soil and surface waters through runoff from agricultural and urban landscapes. POPs are therefore found all around the globe far from their source and bio-concentrate as they move up through food chains and accumulate in fatty tissues of living organisms at higher trophic levels. These chemicals have been banned by many countries however they remain stockpiled, are produced or used illegally, or because they have long half-lives which enable them to continue to exist in soil and other environmental media (Jones and de Voogt 1999; Harrad 2001; Xu et al. 2013).

POPs were widely produced and used during the huge increase of industrial production after the Second World War. Thousands of synthetic chemicals were introduced for commercial use and many proved to be beneficial in pest and disease control, crop production and industries. Unforeseen environmental and human health effects were however accompanied with these intentionally or unintentionally produced chemicals. There are 12 main POPs, referred to as the 'dirty dozen' and include the following (Table 5.1).

Therefore, due to POPs having the ability to bio-accumulate and magnify within food chains, the main concern is usually placed on the impact on top predator species which includes humans. Dioxins, furans and PCBs can be described as by-products of industrial processes which enter the environment through their use as well as disposal. These materials have become a real emerging threat as they have possible long-term negative effects on freshwater as well as other ecosystems. PCBs contamination has become to receive more attention as the contamination has been established to be worldwide. POPs, PCBs as well as other mentioned toxins are therefore a real threat

Table 5.1 The 'Dirty Dozen' POPs, their uses, main characteristics and effects on the environmental and human health (Bluevoice.org 2007)

POP	Use/s	Characteristics and effects/impacts
Aldrin	Organochlorine insecticide: used to control soil insects	• Resistant to leaching into groundwater and is released from soil by volatilisation • Bio-concentrate • Toxic to humans and is a carcinogen and mutagen
Chlordane	A manufactured pesticide used on agricultural crops and extensively in the control of termites	• Binds strongly to soil particles on the surface and not likely to enter groundwater. Can stay in the soil for 20 years • Bio-concentrates in fish, mammals and birds • Damages nervous and digestive systems as well as liver in animals and humans. Has caused convulsions and death • Has been linked to prostate and breast cancers
DDT	Synthetic pesticide	• Toxicant with a half-life of 2–15 years and immobile in most soils • Banned for agriculture worldwide but is still used to some extent in mosquito control in some parts of the world • Bio-concentrates significantly in fish and other aquatic species • Individuals consuming contaminated fish at a chronic level had an increased risk of diabetes occurrences • Probable human carcinogen
Dieldrin	Insecticide	• Closely related to Aldrin (breaks down into dieldrin) • Accumulates in the food chain • Long-term exposure is toxic to humans and animals • Mostly banned across the world • Linked to Parkinson's disease, breast cancer as well as immune, reproductive and nervous system damage
Endrin	Insecticide: used on cotton, maize and rice Rodenticide: used to control mice and voles	• Banned in most countries • Absorbs into sediments and surface water • Bio-concentrates into fatty tissues of organisms living in water • Toxic to aquatic organisms such as fish, aquatic invertebrates and phytoplankton • Endrin poisoning in humans affect nervous system

(continued)

5.2 Water Quality Challenges

Table 5.1 (continued)

POP	Use/s	Characteristics and effects/impacts
Heptachlor	Insecticide	• Similar to Chlordane • Remains in the environment for decades • Possible human carcinogen
Hexachlorobenzene (HCB)	Fungicide: formerly used as seed treatment especially on wheat	• Extremely toxic to aquatic animals. Risk of bio-accumulation in aquatic species is high • Known animal carcinogen (liver, kidney and thyroid) and probable human carcinogen • Can cause liver disease, skin lesions, ulceration, hair loss and thyroid damage in humans • Can accumulate in foetal tissues and transferred into breast milk of animals and humans
Mirex	Insecticide: used to control ants Flame retardant in plastic, rubber, paint, paper and electronics	• Persistent, accumulative and toxic pollutant • Evidence of accumulation in aquatic and terrestrial food chains to harmful levels • Toxic to range of aquatic organisms • Mostly affects animal's liver and is transported across the placenta and passed from mother to child through breast milk • Carcinogen risk to humans • Long-term hazard for the environment
Polychlorinated biphenyls (PCBs)	Used as coolant/insulating fluids; flexible PVC coatings of electric wiring and components; pesticide extenders; cutting oils; hydraulic fluids; sealants; adhesives; wood floor finishes; paints and carbonless copy paper	• Stable compound that does not degrade readily • Detected globally in the atmosphere in most urban areas to regions north of the Arctic Circle carried by wind currents • Bio-concentrates in animals and endocrine disrupting chemical • Transmitted to children through breast milk and placenta • Can cause liver disease, ocular legions, lessened immune responses and can contribute to reproductive problems in humans • Can cause liver, stomach and thyroid damage as well as immune system changes, behavioural changes and impaired reproduction in animals • Probable carcinogen risk for humans

(continued)

Table 5.1 (continued)

POP	Use/s	Characteristics and effects/impacts
PCDDs (Polychlorinated dibenzodioxins) or Dioxins		• Estimated half-life ranges from 7–132 years • 80% of dioxins are emitted by coal burning plants, municipal waste incinerators, metal smelting, diesel trucks, land application of sewage sludge, burning treated wood and trash burn barrels. Also generated in bleaching fibres for paper and textiles • Bio-accumulates in humans and animals. Small amounts in contaminated water can bio-concentrate up into the food chain to dangerous levels • Enters the general population through ingestion of food especially through fish, meat and dairy products • Teratogens (cause birth defects), mutagens and potential human carcinogen • In humans it can have effects on reproductive/sexual development, immune system damage, thyroid and nervous system disorders, endometriosis and diabetes • In animals and fish it can cause cancer birth defects, liver damage, endocrine damage and immune system suppression
Polychlorinated dibenzofurans		• Highly toxic • Similar properties and chemical structure to dioxins
Toxaphene	Insecticide	• Highly toxic • Can cause lung damage, damage to the nervous system as well as kidneys and can be fatal

towards current and future health of a variety of ecosystems as well as a major risk for animal and human health across the globe (Harrad 2001; Xu et al. 2013).

Emerging Contaminants

Most focus has always been placed on identifying the cause and consequences of the following chemicals which degrade the environment namely nutrients, heavy metals, active ingredients in pesticides and POPs. The environmental risks related to emerging contaminants have grown to an increasing concern especially in the past decade. Emerging contaminants originate from a variety of product types which include but are not limited to naturally produced compounds (toxins produced by fungi, bacteria and plants), human pharmaceuticals, veterinary medicines (antibiotics and antiparasitic agents), nanomaterials, bio-terrorism or sabotage agents, human personal care products (essential oils, herbal medicines, antibacterials and fragrances), hormones (synthetic and natural estrogens and androgens) as well as paint coatings. Emerging contaminants may also include endocrine disruptors, pharmaceuticals as well as personal care products which may not be removed by existing wastewater treatment operations also end up in freshwater ecosystems, may impair reproductive success of birds as well as fish and have impacts which are yet to be detected. Certain emerging contaminants may be natural toxins as well as degradation products of man-made chemicals which can be formed in the natural environment by animals, plants and microbes (UNEP 2010).

It has been estimated that 700 new chemicals are introduced into commerce annually only in the USA and worldwide pesticide application is estimated at over 2 million metric tonnes (PAN 2009; Stephenson 2009). The prevalence, transport as well as fate of these new chemicals are largely unknown until recently. It should be noted that emerging contaminants are not necessarily newly discovered chemicals but rather substances which have been present in the environment for a long time but whose presence and significance are only being recognised now. Data regarding emerging contaminants are scarce and detection methods in the natural environment may be non-existent or at an early stage of development. Therefore, the amount of emerging contaminants in water has increased recently due to the development of appropriate detection methods or new testing techniques which allow for the detection of these contaminants at lower and lower levels as well as due to new chemicals being introduced for agricultural, industrial as well as household use which enter and persist in water resources and the environment as a whole.

Endocrine disruptors or synthetic chemicals is a good example of how the threats and impacts of emerging contaminants on the quality of water, human health and the environment as a whole is still not fully understood. These chemicals can interfere with hormone action and many of these endocrine disruptors mimic or block other hormones in the body, disrupting the development of the endocrine system as well as organs which respond to endocrine signals in organisms indirectly exposed during their early development stages. The effects of these chemicals are permanent and can not be reversed (Colborn et al. 1993). Effects on wildlife include thinning of bird eggshells, inadequate parental behaviour as well as cancerous growths to name a few. The effects on humans and human development are less known, however, animal studies have suggested that there is a cause of concern even at low doses. It has also been established that effects may extend beyond the exposed individual by affecting foetuses of exposed pregnant women as well as breastfed children. These chemicals may also have multigenerational effects through the modification of genetic material and heritable mechanisms.

There is an increasing concern regarding pharmaceuticals and personal care products which enter the environment and waterways due to wastewater facilities not being equipped to remove them (Carr and Neary 2008). Low concentrations currently present in waterways have not presented acute health effects but have presented subtle behavioural and reproductive problems for both humans and wildlife as well as likely synergistic impacts when combined with endocrine disruptors. Further research is required to address uncertainties related to their possible impacts or effects.

Lastly, emerging pathogens have also been identified as a possible threat as they appear in human populations for the first time or have occurred before but are increasing in incidence or have been found to expand into areas where they have not been reported before (WHO 2003). These pathogens can emerge due to new environments or changes in environmental conditions, from the use of new technologies as well as from scientific advancements in terms of inappropriate use of antibiotics, insecticides and pesticides which create resistant pathogen strains. The recording of new emerging pathogens or the increase in their incidence have increased in recent decades and have been found to also be a threatening factor towards water quality which needs more research and public attention.

5.3 World's Top Pollutants, Toxic Threats and Dirtiest Places

The world's freshwater resources are facing significant pollution problems, some areas more than others. Most regions across the world are affected by water pollution, causing the degradation of water quality to be a global issue and concern. The biggest water polluting countries of the world include China, USA, India, Japan, Germany, Indonesia as well as Brazil.

It should be noted that the types of pollution which affects developed and developing regions differ in various ways and should, therefore, be taken into account when looking at water pollution issues. The main sources for water degradation in developed countries are established to be agriculture, industries or factories as well as water pollution from fuel emissions especially in large metropolitan areas where human populations are growing at a constant pace. In terms of developing countries, the sources of water pollution differ somewhat, however, the end result is relatively the same. The biggest culprit of water pollution in developing countries includes the lack of or inadequate waste management as well as the primarily the lack of sewerage and septic systems. Agricultural activities also contribute to water pollution as the case in developed countries however the lack of dedicated water sources is an additional issue in developing countries. The lack of dedicated water sources has led to rural areas not having proper water supply infrastructure and in turn forces communities to rely on a single source of surface water or in some cases wells which can be easily polluted. These differences cause certain areas to have 'worse' water pollution than other areas. This should be kept in mind when looking at worst pollution problems and toxic threats as well as the dirtiest places in the world to establish which areas are significant risks to current and future sustainability as well as whether these affected areas would be able to address, mitigate or reduce these significant pollution issues or whether these issues will persist and become an even bigger risk by affecting other areas as well.

5.3.1 Worst Toxic Pollution Problems and Threats

Synthetic and toxic chemicals are released into the environment from different sources on a daily basis and affect the environment as well as human health as a whole. Toxic pollution problems mainly affect the human population who are located within close proximity of the source of pollution which may have adverse effects on human and animal health as well as the environment in terms of water, land and air. The top ten toxic pollution problems found across the globe include the following:

- Lead-acid battery recycling;
- Mercury and lead pollution from mining;
- Coal mining in terms of sulphur dioxide and mercury pollution;
- Artisanal gold mining in terms of mercury pollution;
- Lead smelting;
- Pesticide pollution primarily from agriculture and storage;
- Arsenic pollution in groundwater;
- Industrial waste waters;
- Chromium pollution primarily from the dye industry; and
- Chromium pollution from tanneries (Blacksmith Institute 2007).

Some of these primary toxic pollution problems, in turn, create some of the world's worst pollution problems which have been put together as a list of the new top six toxic threats by Pure Earth and Green Cross Switzerland in 2015 to create awareness on these issues. The top six pollutants which pose the greatest threat to human health together affect the health of 95 million people and account for over 14.7 million disability-adjusted life years (lost years of 'healthy' life) in low and middle-income countries, resulting in debilitating, life-threatening diseases especially in children (Pure Earth and Green Cross 2015). The top six toxic pollutants have been established to be lead, radionuclides, mercury, hexavalent chromium, pesticides as well as cadmium which will each be discussed briefly.

Lead
Lead, a heavy metal and neurotoxin which occurs naturally in the environment, has been mined for centuries for use in numerous products and combined with other metals to form alloys. Sources of lead pollution include mining, smelting as well as lead recycling processes which have effects on both environmental and human health. Exposure includes inhalation of contaminated dust, ingestion of contaminated soil, water or food and dermal contact. Human health effects include neurological damage, IQ decrement, anaemia, nerve disorders as well as a host for other health problems including death.

It is estimated that 26 million people are at risk globally of lead exposure with estimated burden of disease of 9 million. Nearly 800 sites have been identified over the globe where lead threatens health of a population. Most of these identified toxic sites have been caused by used lead-acid battery (ULAB) recycling or smelting and metal extraction activities. ULAB recycling is a major source of recycled lead due to the global demand for lead-acid batteries remains high over the globe. Even though lead is recycled at the highest rate in the world and consequently reduces the rate of land filling and the need to mine for lead, the dangers and risks of unregulated lead recycling remains substantial and is increasing as more populations might become exposed.

Smelting can be described as the application of heat to mineral-rich rock or scrap to extract valuable metals which in turn releases large amounts of pollution into the environment by the related industries and contributes heavily to global lead emissions as well as arsenic, cadmium and chromium emissions around the world. Contaminated effluents and solid waste are also produced and some processes may produce large amounts of sulfuric acid which may contaminate surrounding water bodies (Pure Earth and Green Cross 2015).

Radionuclides

Radionuclides can occur either naturally in soil and rocks such as uranium, thorium and potassium or be artificially produced such as the case of plutonium and americium. Environmental releases are attributed to industrial processes which can include uranium mining, mine waste disposal, nuclear energy production as well as creation of radiological products used in medicine. Exposure can include ingestion and inhalation resulting in acute health effects (nausea, vomiting and headaches) and chronic effects (fatigue, fever, hair loss, diarrhoea, low blood pressure) as well as death. Ionising radiation can also cause cell damage which may result in cancer development or genetic aberration.

It is estimated that between 800,000 and 1 million people are at risk of radionuclide exposure and 91 sites have been identified involving the mining and processing of uranium across the globe which is a major threat to the surrounding human population in terms of health.

Uranium mining and radioactive waste disposal activities are the primary identified toxic sites. Due to limitations in the mining process as well as limited locations of uranium deposits around the globe, uranium extraction only occurs in selected countries. Most common means of obtaining uranium includes open-pit mining and in situ leaching, which result in exposed uranium and hazardous tailings materials contaminating the surrounding area and pose major risks to workers and nearby populations. Waste from uranium mining includes toxic combinations of heavy metals, chemical agents from the leaching process as well as contaminants such as arsenic. Radioactive waste which is a by-product of nuclear reactors can also be a source of exposure to uranium, but can also come from fuel processing plants, power generation facilities, military exploits, hospitals and medical research facilities (Pure Earth and Green Cross 2015).

Mercury

Mercury can occur as a heavy metal as elemental mercury which is used in variety of industrial processes and found in products such as thermometers and dental fillings. Exposure to elemental mercury can cause brain, kidney and immune system damage and impair foetal development. Mercury can also occur as a chemical compound in the form of inorganic mercury, found in numerous industrial compounds and produced when elemental mercury combines with carbon to form methylmercury (potent neurotoxin). Exposure to harmful levels of organic mercury is unlikely however methylmercury can cause Minamata disease (severe neurological syndrome).

It is estimated that 19 million people around the world are at risk of mercury exposure with a burden disease of approximately 1.5 million. Over 450 toxic sites have been identified around the globe threatening the health of surrounding populations. The main pollution problem leading to increased exposure of mercury around the globe can be attributed to artisanal small-scale gold mining (ASGM) which is primarily small-scale informal operations focused on the extraction and processing of metals. Many of these sites use mercury for gold extraction with little or no safety or environmental controls. The discarded materials from this activity often contain other harmful heavy metals contaminating nearby areas and the burning of the amalgam exposes workers and their families to mercury vapours (Pure Earth and Green Cross 2015).

Hexavalent Chromium

Chromium, a naturally occurring heavy metal, is often used in industrial processes. It can be released into the environment through natural processes but is often released by industrial activities particularly by leather processing. It is found in two forms namely chromium III (trivalent, occurs naturally and most stable) and chromium VI (hexavalent, from industrial processes and accompanied by numerous deleterious health effects). Depending on the type of exposure, chromium can cause respiratory and gastrointestinal system damage. Hexavalent chromium is also known to be a human carcinogen and can increase the rates of some cancers based on the exposure route. Even though trivalent chromium is low in toxicity, it can still cause negative health effects and damage to DNA.

It is estimated that 16 million people are at risk of exposure globally, with an estimated burden disease of 3 million, and over 300 sites have been identified around the world threatening the health of human populations. The biggest pollution culprit related to this toxic problem is tannery operations in the leather industry related to the tanning and product manufacturing sector.

Tanning processes produce large amounts of toxic waste which, in some cases, is discharged into surrounding water sources but may also dump large amounts of chromium-contaminated solid wastes into riverbanks and surrounding areas. Some of these solid wastes can include skins, hides and fats. The overall lack of environmental safety controls as well as environmental monitoring mostly results in the contamination of surrounding water sources and soil surfaces from additional tanning agents and treatment processes. Even though pipelines and canals may exist to enable the transport of waste away from the tanning facilities, these often run through nearby villages and leaks can pose great health hazards and lead to the contamination of agricultural soils, residential spaces and drinking water sources (Pure Earth and Green Cross 2015).

Pesticides

Pesticides can be chemical or heavy metal substances which have been used extensively throughout the world to protect crops by eliminating pests and consequently increased agricultural output or control disease vectors. Significant amounts of pesticides have been washed away by rainfall into surface and ground waters which have exposed surrounding populations. General acute health effects include headaches, nausea as well as convulsions and chronic exposure may lead to neurological, reproductive as well as dermatological health effects.

It is estimated that around 7 million people are at risk globally with an estimated burden of disease of 1 million. More than 200 toxic sites have been identified around the world and the continued global, long-term use of pesticides will increase this number in future. The main pollution problem which contributes to these toxic problem areas is primarily agricultural use thereof as well as production and storage activities. Agriculture being one of the most dominant sectors around the world especially in middle and low-income economies has been found to benefit the income of a country's population by two to four times to that from the GDP growth from other industries. Due to this, pesticides have been used and promoted extensively over the decades to improve agricultural output, enhance crop growth and protect crops from harmful pests. Many of these pesticides pose harmful health risks to both the environment and human population and there is an increased difficulty to balance the need for increased agricultural output or local food security and the desire to protect health of the environment and the human population.

Production and storage contribute to this toxic problem due to the continued production and storage of pesticides labelled as highly hazardous or for restricted use. Some of these can also be waste-mined and sold on black markets. Many uncontrolled sites exist which contain these compounds and there are a high number of stockpiles of these hazardous pesticides especially in low and middle-income countries. Many of the structures built for the Green Revolution are now abandoned and the lack of maintenance and oversight of these containment facilities have consequently contaminated the surrounding environment due to leaks and cracks. In most cases, these pesticides have been stored in drums which are open, deteriorated or leaking. Waste-mining usually takes place at these locations by small businessmen who repackage these into small bags and sell them in local markets (Pure Earth and Green Cross 2015).

Cadmium

Cadmium, a by-product of zinc mining and processing, is released into the environment by smelting and mining wastes. It also occurs in some phosphate rock and in some fertilisers and has a limited industrial use in electroplating and batteries. It can be transported long distances from smelters as small airborne particles and deposited on soil which is consequently taken up into the food chain by plants and can also dissolve in water affecting fish.

Humans therefore get exposed through a contaminated food chain or from active or passive smoking. Cadmium can be harmful in small amounts and can cause significant health impacts for the surrounding population. It is estimated that 5 million people are at risk of exposure around the globe with an estimated burden of disease of 250,000. Approximately 150 sites have been identified around the world where the exposure threatens the health of the population (Pure Earth and Green Cross 2015).

Table 5.2 Top ten most polluted countries and rivers across the globe (Schiller 2013)

	Top ten most polluted	
	Countries	Rivers
1	China	Ganges River—India
2	India	Yellow River—China
3	USA	Doce River—South East Brazil
4	Bangladesh	Citarum River—West Java, Indonesia
5	Nigeria	Mississippi River—Widespread in the USA
6	United Arab Emirates	Sarno River—Southern Italy
7	Pakistan	Marilao River—Bulacan Province, Philippines
8	Mongolia	Buriganga River (Old Ganges)—Bangladesh
9	Egypt	Cuyahoga River—Cleveland, Ohio, USA
10	Iran	Matanza-Riachuelo River—Buenos Aires Province in Argentina

5.3.2 Most Polluted Countries and Rivers

The world's surface and groundwater resources as well as its environment and human population as a whole are therefore threatened by numerous types of contaminants as well as toxic pollution problems which may have immediate as well as significant long-term effects. The world's freshwater resources are faced with increasing pollution from the main water sectors and some areas already can not cope with the amount of waste which has been pumped into its water resources. The world has consequently become a polluted place mainly due to the constant growth in agriculture and industrialisation needed to support the continued exponential growth in the human population. The top ten most polluted countries and rivers have been listed as the following (Table 5.2).

Most countries around the world do not have the necessary environmental controls to try and mitigate degradation and the clean-up process (if any) is usually very slow and expensive, taking place over a period of years to decades. Clean-up processes are however needed in polluted areas to avoid them becoming uninhabitable. Short case studies of the top three most polluted countries and rivers in the world follows.

USA: Insatiable Appetite and Vast Carbon Footprint
One might think it is surprising to find the USA even on this list and in the top three as its lakes and rivers are relatively clean. This is however not the case as it is estimated that the majority of its rivers and streams are unable to support aquatic life and the trend is expanding to other areas across the country. Approximately 55% of its waterways are labelled as being in poor condition and 23% fair. Only 21% are estimated to be of good condition able to support healthy biological communities (Bennet 2013).

The continued degradation is attributed to a variety of factors. The constant increase in food consumption leads to an increase in the use of pesticides and fertilisers by the agricultural sector to try and meet the growing demand which consequently causes nutrient pollution. The continued increase in the country's human population is also accompanied with increase in industrial activities (production of materials or product and energy) which leads to increase in waste waters entering surrounding surface waters through runoff or as point pollution. It was also estimated by the country's Department of Energy that the average American adds 17.3 tonnes of carbon dioxide into the atmosphere per annum which is the highest per capita by any large country. The increased carbon footprint as well as the increase of human-produced chemicals in waterways affects the country's ecological and human health. The country dumps huge amounts of chemical fertiliser into its waterways, choking healthy plant life and produces dead zones (Bennet 2013; Fortin 2017).

Pathogen pollution from animal as well as human waste also contaminates surrounding waterways as well as groundwater. In terms of pollution of waterways by human waste, septic systems have been found to be failing across the country which has led to untreated waste materials freely flowing into surrounding waterways. An estimated 1.2 trillion gal of storm water, untreated sewage as well as industrial waste is discharged into the country's waters annually which has in some areas compromised drinking water. Flint, Michigan is a good example of this where it was found that the levels of lead in water supply is too high and toxic for human consumption due to lead corrosion taking place throughout the city's pipelines which has still not been cleaned up. Other examples of contaminated drinking water in the country include West Virginia where the Elk River chemical spill contaminated downstream cities such as Charleston and nine other counties. It was estimated in 2015 that approximately 77 million people in the country lived in areas where water systems were in some violation of safety regulations and that drinking water contained high levels of lead, nitrates, arsenic or other pollutants. This issue is however not new as these safety violations have been reported multiple times and repercussions for these violations have been described as non-existent where nine out of ten violations are subject to no formal action. The lack of repercussions has been attributed to the complex regulatory system where the responsibility to monitor adherence to federal laws fall on the state.

Therefore, even though the country can afford to clean-up its own waste and invest in the upgrading, maintenance or development of water treatment infrastructure, the lack of repercussions, inadequate investment in infrastructure and continued focus on socio-economic growth has caused continued degradation of the country's water resources leading to both environmental destruction with the development of dead zones as well as the contamination of drinking water with significant human health effects (Gusovsky 2016).

India: Too Many People
India is one of the world's most populous countries (second highest) and was declared in 2011 by the World Economic Forum to have the worst air quality in the world. The poor air quality is mainly attributed to the huge amount of the population driving

5.3 World's Top Pollutants, Toxic Threats and Dirtiest Places

vehicles which are heavy polluters. The country is also the third largest carbon emitter mainly due to the use of coal to generate most of its electrical power. It is estimated that the poor air quality kills 2 million people per annum.

Water pollution is also a serious problem and its rivers have been described to be running sewers. Approximately 70% of the country's surface waters and a growing percentage of its groundwater reserves are being depleted and contaminated by toxic, organic, inorganic and biological pollutants with many of its water resources been rendered as unsafe for human consumption and other activities such as irrigation and industries. This has in turn led to water scarcity within some regions due to the water not being fit for human use and the ecosystem (Wheeler 2016).

Water quality monitoring results for the country indicates that organic and bacterial contamination was critical. Sources of these critical water quality levels include the discharge of domestic and industrial wastewater, primarily untreated wastes from urban centres. Other important reasons for the continued water degradation in the country include the following:

- Improper agricultural practices;
- Depletion of water quantities in rivers and plains;
- Domestic and Industrial waste;
- Social and religious practices such as the dumping of dead bodies, bathing etc.;
- Inadequate treatment of domestic and industrial waste waters;
- Acid rain, eutrophication and denitrification; and
- Climate change.

The uncontrolled rate of urbanisation in the country has been labelled as the primary reason for water pollution in the region as it has led to multiple long-term environmental issues such as the lack of water supply and the generation and collection of wastewater. Uncontrolled urbanisation has led to an increase in the generation of sewage water as well as other domestic and industrial waste. It is estimated that urban centres produce close to 40,000 million L of sewage on a daily basis and barely 20% is treated. The treatment and disposal of wastewater is also a major issue and has consequently intensified the country's water degradation problems in terms of surface and groundwater (AFP 2013; Sumadranil 2016).

The vast amounts of water pollution have led to some adverse effects on both environmental and human health in the vicinity of the polluted water body or the use thereof. Pollution can also be detrimental for the agricultural sector by reducing fertility of soil and crop production. The pollution of the ocean also affects oceanic life in a negative way however the most detrimental effects of polluted water has been on the health of the country's human population which is already low especially in rural areas. Primary ailments caused by water pollution include water-related diseases such as cholera, tuberculosis, dysentery, jaundice, diarrhoea and 80% of stomach ailments are attributed to the consumption of polluted water.

It has been estimated that environmental factors contribute to 60 years of ill-health per 1,000 population in India, compared to 54 in Russia, 37 in Brazil and 34 in China. The socio-economic costs of water pollution have been calculated to include the death of 1.5 million children under the age of five per annum due to water-related diseases, the loss of 200 million person days of work per annum and the country is estimated to lose Rs. 366 billion per annum due to water-related diseases (Parikh 2004). Furthermore, the country as a whole loses 90 million days a year due to water-related diseases with production losses and treatment costs worth Rs. 6 billion. The country also experiences a loss of 30.5 million disabilities adjusted life years due to poor water quality, sanitation as well as hygiene.

In terms of groundwater, these resources are vastly contaminated with fluoride as well as arsenic. The high concentrations of fluoride in drinking water may cause fluorosis which results in weak bones, weak teeth and anaemia. Arsenic has been linked to cancer and has caused health risks for 35–70 million people in West Bengal, Bihar as well as Bangladesh. The state of Bihar is a prime example of the presence of arsenic in water. Bihar is one of the country's most impoverished states where more than 10 million people are estimated to be threatened by arsenic poisoning or arsenicosis from contaminated groundwater resources. Chronic arsenic poisoning can lead to cancer. Various forms of cancer have been attributed to arsenic and include primarily skin cancer followed by other cancers such as bladder, kidney and lung (Arora 2017).

The cost of industrial water pollution abatement has been estimated to be 2.5% of the industrial GDP of the country (Murty and Kumar 2011) and the cost of avoidance is much lower than the damage costs (Parikh 2004).

The weak or non-existent enforcement of environmental legislation, rapid urban expansion as well as the lack of awareness related to the dangers/risks of sewage are blamed for the continued degradation.

China: Supreme Industrial Power leading to Immense Pollution

China has become one of the greatest industrial powers and is one of the world's most populous countries. It has become the world's worst polluter in terms of carbon emissions and polluted its land, water and air in its mission to become the world's supreme industrial power. China is legendary for its air pollution which has been specially called the 'smogpocolypse' even though it has made some successful attempts to improve its poor air quality.

Even though some of the smog has lifted slightly in recent years, the country's water pollution crisis continues to grow with 85% of water in cities major rivers being not suitable for drinking and 56.4% unfit for any purpose in 2015. Only 39.9% of water in Beijing was deemed as functionless and in Tianjin, principle port city with a population of 15 million in northern China, only 4.9% of water was suitable for drinking. The most recent state of the environment report for the country has reported the following in terms of the country's overall water quality:

- A proportion of 'very bad' groundwater quality has decreased slightly.
- The overall proportion of 71.2% of the country's seven major rivers falls between Grade I to III with five rivers being below this average.
- Despite some improvements in key lakes and reservoirs, the proportion of Grade IV and V have risen (MEP 2017).

80% of groundwater abstracted and used by agriculture, industries as well as households across the country is unfit for drinking or bathing due to contamination by industry and farming activities. Approximately 32% of wells located in northern and central China have Grade IV quality water which means it is only fit for industrial uses. Additional 47.3% were at Grade V. Main contaminants of underground water include manganese, fluoride and triazole (fungicide compounds) and in some cases heavy metals. The contamination of underground water located near the surface has forced more cities to dig deeper underground to find clean water and in turn places increased pressure on deep aquifers. Surface waters are polluted primarily by industrial and domestic wastewaters. Approximately one-third of industrial wastewater and 90% of household sewage is released into rivers and lakes untreated. An estimated 80% (278 cities) of the country's cities do not have sewage treatment facilities and few have plans to construct any (Hays 2014; Buckley and Piao 2016).

The water shortages and pollution within the country have escalated to such an extent that the World Bank has issued a warning of 'catastrophic consequences for future generations'. Only half of the country's population has safe drinking water and two-thirds of the rural population use water which is contaminated by human and industrial waste. The water consumed in the country contains dangerous levels of arsenic, fluorine as well as sulphates and it is estimated that 980 million of the 1.3 billion people drink partially polluted water daily. More than 600 million people drink water which has been contaminated by animal and human waste and 20 million people consume well water which is contaminated with high levels of radiation. High levels of arsenic have been measured and have been linked to the high rates of liver, stomach and esophageal cancer. Northern China is faced with more serious challenges where 45% of water is unfit for human consumption in comparison to 10% in southern China. Lastly, 80% of rivers in the northern region have been rated unfit for human contact which highlights the serious problem of water pollution and consequent water shortages in the region (Hays 2014).

The possible reasons for the continued rate of water pollution within the country may include the lack of local governments of enforcing environmental legislation especially on polluting industries as well as insufficient nationwide standards for the treatment of sewage (Tingting 2017). China has to invest an estimated $215 million annually to improve its air quality but must first convert all coal-burning power plants to natural gas or nuclear, remove all badly polluting old cars and trucks as well as develop more low-carbon polluting energy sources. In terms of its water pollution problem, the country has not made major improvements in the last decade which has led to the continued decrease in water quality across the country with major environmental consequences, human health effects and risks as well as increases the risk of water scarcity across the region.

Doce River: From 'Sweet Water' to Sludge
The Doce River runs through the southeast of Brazil and has been called 'sweet water' as it provides much-needed freshwater to the largest steel making region in Latin America. The river has however been listed as the third dirtiest river in the world due to one of the worst environmental disasters experienced by the country in 2015. Two containment dams ruptured, spilling 60 million m^3 of iron ore sludge into the river, killing 17 people, injuring scores of others and leaving the river to be mud. The

dumped sludge was mostly made up of heavy metals which consequently destroyed the river's aquatic life and devastating the livelihoods of numerous fishermen as well as drinking water for many people. Many communities along the river have suffered from diarrhoea as well as vomiting as the toxic mud started to contaminate the water supply. The metallic dust from the river will also pose air borne health risks (Bevins 2015; Reuters 2015).

An estimated 50 million tonne of iron ore and toxic waste was dumped into the river that day with sludge covering riverbanks as well as croplands along the entire length of the river (853 km). The environmental consequences included complete devastation, killing fish as well as other wildlife. The drinking water supply for most of the river valley has been contaminated and the toxic sludge has reached the Atlantic Ocean. Eleven of the 90 native fish species in the river was already at risk of extinction before the spill and it is believed that wide-ranging forms of fauna and flora (some yet to have been discovered) will be wiped out as entire ecosystems have been destroyed. It is estimated that surviving ecosystems could take from 10 to 50 years to regenerate but will not be the same as before (Bevins 2015).

Unacceptable levels of arsenic as well as mercury were measured days after the spill which led to numerous human fatalities and flooding thick mud across two Brazilian states (Reuters 2015). Eight months after the spill, approximately 1.6 million people living along the river were still struggling with health risks as well as with a crisis of public confidence due to contaminated drinking water. The indigenous Krenak community were severely affected as the spillage destroyed not only their water supply but harmed their culture as it is sacred to them forming part of their culture and their lives. The river was also their primary food as well as drinking water source. The community is now dependent on the supply of water trucks by Samarco and have complained about this water having high levels of chlorine which irritate their skin and stomachs. Authorities and Samarco, however, say that the water is indeed safe for drinking as it comes out of their river water treatment facility. People have become sick and a public confidence issue has developed as a result (Yeomans and Bowater 2016).

BHP Billiton, who built the burst tailings dams, has been sued by the Brazilian government for $5 billion but whether the Doce River will ever be cleaned up after one of the world's worst environmental disasters in history remains to be seen.

Yellow River: Toxic Waste Dump

As indicated previously, China has transformed from an impoverished farming-reliant country into an industrial power in the last three decades with major environmental costs. The rivers and groundwater upon which its population depends on have been poisoned through the widespread dumping of chemicals as well as industrial wastewater. The Yellow River, which is deemed as the country's 'mother river' has been severely overexploited as some parts of it has run dry while other parts are polluted and groundwater aquifers are severely under pressure. Main source of pollution are thousands of petrochemical factories which are located along both the Yellow and Yangtze rivers which freely dump their wastewater specifically into the

Yangtze River, threatening the lives and health of many communities (Hays 2013; Berlin 2017).

The Yellow River is regarded by some to be the world's muddiest major river as it discharges three times the sediment of the Mississippi River and gets its name and colour from the yellow silt it picks up in the Shaanxi Loess Plateau. This river is vital in making the northern parts of China habitable, supplies water to 155 million people (12% of the population) and irrigates 15% of the country's farmland. The river also travels through industrial centres, the major coal-producing region and huge population centres. Approximately 4,000 of China's 20,000 petrochemical factories are located on the Yellow River. A third of all fish species found in the river have become extinct due to the construction of dams, falling water levels as well as pollution and overfishing. More than 80% of the river basin is chronically polluted mainly by 4 billion tonnes of waste water which flows into it annually. Canals which were once filled with fish have changed to the colour purple from red wastewater from numerous chemical plants causing the death of aquatic ecosystems as well as causing the water to be too toxic for drinking or for agricultural use. Even though the Yellow River is essential for the well-being of the country, sewage and industrial wastewater are continually dumped into the river which has made it a toxic waste dump (Hays 2013; Berlin 2017).

Some examples of pollution include the leaking of 6 tonnes of diesel oil in December 2005, into a tributary of the river from a cracked pipe due to freezing conditions, produced a 40 mile long slick and resulted in 63 water pumps having to be shut down in Jinan. Furthermore, in October 2006, a one-kilometre section of the river turned the colour red due to red and smelly discharge from a sewage pipe in the city of Lanzhou. An estimated 50% of the river has been designated as biologically dead and some areas along the river have recorded dramatic increases in cancer, birth defects as well as waterborne diseases. The rates of cancer have increased so dramatically in some villages that they have now been designated as cancer villages which include for example the Xiaojidian village, located in Shandong on a tributary of the river. Wastewater from tanneries, paper mills as well as factories have been blamed as the primary sources of pollution and have caused 70 people to die from stomach cancer in a five year period in a village of only 1,300 people. Additionally, more than a thousand other people have died in the other 16 surrounding villages.

China has numerous environmental regulations however these are seldom prioritised or enforced. Officials are rather rewarded for economically advancing their areas than ensuring environmental compliance and receive a huge amount of incentives to place short-term economic gains ahead of long-term environmental goals. There are however environmental activists in the country who have made it their mission to clean-up the Yellow River by improving declining ecosystems in western China. The Green Camel Bell group are dedicated to cleaning up areas but will do little more than educate people (Berlin 2017).

The responsibility once again falls on the Chinese government to stop urban areas and industries from dumping wastes into the river. Some drastic measures have however been adopted and have become stricter in the past couple of years. In 2011, three major urban centres which included Baoji, Xianyang, and Xi'an, were fined a total of US$13 million for excessive pollution and more than a hundred polluting factories along the river have been shut down in the past decade. This has resulted in some improvement in the river's water with some areas experiencing a returning of the famous yellow colour as well as a decrease in pollution indicators such as ammonia nitrogen density especially in Tongguan, Shaanxi Province (Hong'e 2017).

Ganges River: Sacred River used as Human Dumping Ground
Lastly, the Ganges River, the most sacred river in Hinduism and third largest river in the world by discharge is also the most polluted river in the world. The river affects the lives of approximately 400 million people who live near it who unfortunately dump their waste into the river as they use it for drinking, bathing and cooking which in turn gives rise to waterborne illnesses.

The Ganges River covers 29 cities having a population of more than 100,000, flows through 23 cities having population between 50,000 and 100,000 and near about 48 towns, emphasising its importance to numerous communities within the country. The river consequently tolerates domestic wastewater, untreated industrial waste as well as religious events causing the river to have pollutants at levels more than 3,000 the permissible limit. The primary source of pollution has been labelled to the direct pouring of human faeces, urine as well as sewage into the river as well as the throwing of half-burned or unburned human dead bodies and animal carcasses into the river hoping to purify their souls and be immediately received by heaven. Runoffs of harmful pesticides as well as fertilisers from agricultural activities also play a role in the degradation. Mercury has also been found in the river however it has not yet reached alarming levels (Trivedi 2010; Dhillon 2014).

Raw human and animal sewage, plastic bags and bottles, industrial effluents, chemicals from tanneries, partially cremated corpses, garlands of flowers, human remains, animal carcasses, butcher's offal, chemical dyes from sari factories and construction waste have all caused the Ganges River to become the most polluted river in the world and become unfit for drinking, bathing as well as agricultural purposes. It is estimated that more than 3,000 million L of untreated sewage is dumped into the river on a daily basis and toxins such as chromium are introduced into the river by leather factories. Water-related diseases in the river are at an all-time high and it has become the leading cause of infant/child mortality rates, skin problems as well as other serious disabilities (Trivedi 2010; Dhillon 2014).

Not only human life has been affected by the pollution. The Yamuna River, tributary of the Ganges, has been devoid of marine life for the past 15 years and the country's whole fishing industry has completely collapsed. Coral as well as fish have been killed in large amounts from the heavily polluted water which is not able to support any life.

The Ganges River not only experiences the dumping of chemical waste but its water supply is also being drained by factories which decreases the amount of water available for the population. Large amounts of water are being drained from the river and factories will often also divert rivers to meet their water demand. This, in turn, deprives surrounding communities living off the water and causes droughts. An example of such a factory has been Coca-Cola which has been accused of diverting rivers and consequently damaging the environment and destroying entire ecosystems several years ago. The vast irrigation networks which have been developed in an effort to try and feed the country's expanding population has also contributed to a decrease in water levels especially in the dry season (Trivedi 2010; Dhillon 2014).

Attempts have however been made to try and clean-up the Ganges River by organisations such as the Ganga Action Plan, which began in 1985. This project has, however, become a dismal failure due to being vastly underfunded but it is still a start. In the view of the ever increasing water scarcity in the basin, it is of great importance that no wastewater be discharged into the river and that clean-up operations be properly funded and put into place to ensure future water availability within the region.

5.4 Conclusions

Therefore, further development of areas, regions and the world as a whole has been accompanied with an array of changes which have resulted in numerous changes in the terrestrial component of the water cycle. The global water system is thus not isolated and universal transformations will be accompanied with changes in freshwater systems in terms of alteration of physical, chemical and biological characteristics as well as anthropogenic water use and withdrawal.

The world's water resources are physically changed through long-term changes in surface and subsurface moisture storage and runoff, and persistent changes in precipitation and hydrological patterns. Developments such as mining operations physically alter freshwater systems through altering characteristics such as the system's soils, wetland hydrology and geomorphology within one region and have unintended altering effects or cumulative impacts such as increased sedimentation or the altercation of a different freshwater system in another connected region or area (Alcamo et al. 2008).

The chemical and biological characteristics can be altered through long-term changes in the flow of nutrients and sediments towards the oceans, as well as the key levels of water quality and habitat parameters. The over utilisation of freshwater systems through various human activities may cause an influx of nutrients within aquatic habitats as indicated previously in this chapter and consequently greatly reduce aquatic organisms and hold severe consequences for aquatic ecosystems (Alcamo et al. 2008).

Last, anthropogenic water use and withdrawals alter water resources through rapidly changing patterns of water consumption across different economic sectors and regions. Industrialised countries now tend to be associated with reduced withdrawals of water while the volumes of water withdrawn in the developing regions are increasing. These trends have caused changes in water stress patterns with uncertain global implications (Alcamo et al. 2008).

The world's water resources are therefore constantly being degraded by both natural and human factors through point and non-point sources. Human activities especially in terms of the three major water uses is, however, the greatest source of degradation which has widespread ecological and human health consequences across the globe.

References

Adams SB, Kolo RJ (2006) Public health implications of Gurara River around Izam Environs, Niger State, Nigeria. In: Fisheries Society of Nigeria (FISON) conference proceedings, Calabar, 13–17 Nov 2006, pp 167–173

AFP (Agence France Presse) (2013) India river pollution: 80 percent of Indian sewage flows untreated into country's rivers. Available via https://www.huffingtonpost.com/2013/03/05/india-river-pollution-sewage_n_2810213.html. Accessed on 26 Feb 2018

Alcamo JM, Vorosmarty CJ, Naiman RJ, Lettenmaier DP, Pahl-Wostl C (2008) A grand challenge for freshwater research: understanding the global water system. Environ Res Lett 3:1–6

Arora M (2017) Arsenic-polluted water linked to cancer in India. Available via https://edition.cnn.com/2017/04/28/health/arsenic-water-pollution-cancer-india/index.html. Accessed on 26 Feb 2018

Beaudry F (2017) Water pollution: nutrients. Available via https://www.thoughtco.com/water-pollution-nutrients-1204127?utm_source=emailshare&utm_medium=social&utm_campaign=shareurlbuttons. Accessed on 26 Feb 2018

Bennet D (2013) Half of all U.S. rivers are too polluted for our health. Available via http://www.theatlantic.com/national/archive/2013/03/half-all-us-rivers-are-too-polluted-our-health/316027/?utm_source=eb/. Accessed on 26 Feb 2018

Berlin J (2017) What's killing the Yellow River? Available via https://www.nationalgeographic.com/magazine/2017/07/china-yellow-river-fragile-landscape/. Accessed on 26 Feb 2018

Bevins V (2015) As Brazil mine spill reaches ocean, its catastrophic extent becomes clear. Available from http://www.latimes.com/world/brazil/la-fg-brazil-spill-20151220-story.html. Accessed on 26 Feb 2018

Blacksmith Institute (2007) The world's worst pollution problems: the top ten of the toxic twenty. Available via http://www.worstpolluted.org/. Accessed on 26 Feb 2018

Bluevoice.org (2007) Persistent organic pollutants (POPs). Available via http://www.bluevoice.org/POPsFactSheet.php. Accessed on 26 Feb 2018

Buckley C, Piao V (2016) Rural water, not city smog, may be China's pollution nightmare. Available via https://www.nytimes.com/2016/04/12/world/asia/china-underground-water-pollution.html. Accessed on 26 Feb 2018

Campbell PGC, Stokes PM (1985) Acidification and toxicity of metals to aquatic biota. Can J Fish Aquat Sci 42:2034–2049

Canedo-Arguelles M, Hawkins CP, Kefford BJ, Schafer RB, Dyack BJ, Brucet S, Buchwalter D, Dunlop J, Fror O, Lazorchak J, Coring E, Fernandez HR, Goodfellow W, Achem ALG, Hatfield-Dodds S, Karimov BK, Mensah P, Olson JR, Piscart C, Prat N, Ponsa S, Schulz C-J, Timpano AJ (2016) Saving freshwater from salts. Science 351:914

Carr GM, Neary JP (2008) Water quality for ecosystem and human health, 2nd edn. United Nations environment programme global environment monitoring system. Available via http://www.gemswater.org/publications/pdfs/water_quality_human_health.pdf. Accessed on 26 Feb 2018

Castro J, Reckendorf F (1995) Potential NRCS actions to improve aquatic habitat—working paper no. 6. Available via https://www.nrcs.usda.gov/wps/portal/nrcs/detail/national/technical/?cid=nrcs143_014201. Accessed on 26 Feb 2018

Colborn T, Soto A, vom Saal F (1993) Developmental effects of endocrine-disrupting chemicals in wildlife and humans. Environ Health Perspect 101:378–384

References

Dakkak A (2016) Water pollution worries in developing world. Available via http://www.ecomena.org/water-pollution/. Accessed on 26 Feb 2018

Dhillon A (2014) The Ganges: holy river from hell. Available via http://www.smh.com.au/national/the-ganges-holy-river-from-hell-20140806-100xz9.html. Accessed on 26 Feb 2018

Domske H, O'Neil CR (2003) Invasive species of lakes Erie and Ontario. New York Sea Grant. Available via http://www.seagrant.sunysb.edu/ais/pdfs/AIS-LErieOnt.pdf. Accessed on 26 Feb 2018

EPA (United States Environmental Protection Agency) (2012) The facts about nutrient pollution. EPA-840-F12-003

Eyres W (2009) Water chestnut (*Trapa natans* L.) infestation in the Susquehanna River watershed: population assessment, control, and effects. Biological field station Oneanta N.Y. Occasional paper no. 44

Fortin J (2017) America's tap water: too much contamination, not enough reporting, study finds. Available via https://www.nytimes.com/2017/05/04/us/tapwater-drinking-water-study.html. Accessed on 26 Feb 2018

Gusovsky D (2016) America's water crisis goes beyond Flint, Michigan. Available via https://www.cnbc.com/2016/03/24/americas-water-crisis-goes-beyond-flint-michigan.html. Accessed on 26 Feb 2018

Hamelink JL, Landrum PF, Harold BL, William BH (1994) Bioavailability: physical, chemical, and biological interactions. CRC Press Inc, Boca Raton, FL

Harrad S (2001) Persistent organic pollutants: environmental behaviour and pathways of human exposure. Kluwer, Boston, MA

Hays J (2013) Yellow River: facts and details. Available via http://factsanddetails.com/china/cat15/sub103/item448.html. Accessed on 26 Feb 2018

Hays J (2014) Water pollution in China. Available via http://factsanddetails.com/china/cat10/sub66/item391.html. Accessed on 26 Feb 2018

Hong'e M (2017) Water quality of Yellow River's largest tributary returning to normal. Available via http://www.ecns.cn/2017/03-01/247477.shtml. Accessed on 26 Feb 2018

Hubbard RK, Newton GI, Hill GM (2004) Water quality and grazing animals. J Anim Sci 82:255–263

Jones KC, de Voogt P (1999) Persistent organic pollutants (POPs): state of the science. Environ Pollut 100:209–221

Kabata-Pendia A (2001) Trace elements in soils and plants. CRC Press, Boca Raton, FL

MEP (Ministry of Environmental Protection) (2017) State of the environment report review, 2016

Murty MN, Kumar S (2011) Water pollution in India: an economic appraisal. India infrastructure report. Water: policy and performance for sustainable development

PAN (Pesticide Action Network) (2009) PAN pesticides database. Available via http://www.pesticideinfo.org/. Accessed on 26 Feb 2018

Parikh J (2004) Environmentally sustainable development in India. Available via http://scid.stanford.edu/events/India2004/JParikh.pdf. Accessed on 26 Feb 2018

Pure Earth and Green Cross (2015) World's worst pollution problems: the new top six toxic threats—priority list for remediation. Available via http://www.pureearth.org/release-of-2015-worlds-worst-pollution-problems-report/. Accessed on 26 Feb 2018

RAMP (Regional Aquatics Monitoring Program) (2008) Water quality indicators: temperature and dissolved oxygen. Available via http://www.ramp-alberta.org/river/water+sediment+quality/chemical/temperature+and+dissolved+oxygen.aspx. Accessed on 26 Feb 2018

Reuters (2015) Arsenic and mercury found in river days after Brazil dam burst. Available via https://www.theguardian.com/business/2015/nov/26/brazil-dam-arsenic-mercury-rio-doce-river. Accessed on 26 Feb 2018

Schiller B (2013) The 10 most polluted places in the world. Available via https://www.fastcompany.com/3021425/the-10-most-polluted-places-in-the-world. Accessed on 26 Feb 2018

Stephenson J (2009) Testimony before the subcommittee on commerce, trade, and consumer protection, committee on energy and commerce, House of Representatives: options for enhancing the effectiveness of the toxic substances control act. US Government Accountability Office

Sumadranil (2016) Water pollution in India: causes, effects and solutions. Available via https://www.mapsofindia.com/my-india/education/water-pollution-in-india-causes-effects-solutions. Accessed on 26 Feb 2018

Tchounwou PB, Yedjou CG, Patlolla AK, Sutton DJ (2012) Heavy metals toxicity and the environment, vol 101. NIH-PA, pp 133–164

Tingting D (2017) In China, the water you drink is as dangerous as the air you breathe. Available via https://www.theguardian.com/global-development-professionals-network/2017/jun/02/china-water-dangerous-pollution-greenpeace. Accessed on 26 Feb 2018

Trivedi RC (2010) Water quality of the Ganga River—an overview. Aquat Ecosyst Health Manag 13:347–351

UN WWAP (United Nations World Water Assessment Programme) (2003) United Nations world water assessment programme. The 1st UN world water development report: water for people, water for life. Available via http://www.unesco.org/water/wwap/wwdr/wwdr1/. Accessed on 26 Feb 2018

UNEP (United Nations Environmental Programme) (2010) Clearing the waters. Pacific Institute

UNEP (United Nations Environmental Programme) (2016) A snapshot of the world's water quality: towards a global assessment. United Nations Environment Programme, Nairobi, Kenya, p 162

Verkleji JAS (1993) The effects of heavy metals stress on higher plants and their use as biomonitors. In: Markert B (ed) Plant as bioindicators: indicators of heavy metals in the terrestrial environment. VCH, New York, pp 415–424

Wheeler A (2016) World water day 2016: India has the world's worst water quality, and it's getting worse. Available via http://www.ibtimes.co.uk/worldwaterday2016indiahasworldsworstwaterqualityitsgettingworse1551050. Accessed on 26 Feb 2018

WHO (World Health Organisation) (1996) Trace elements in human nutrition and health. World Health Organization, Switzerland, Geneva

WHO (World Health Organization) (2003) Emerging issues in water and infectious disease. Available via http://www.who.int/water_sanitation_health/emerging/emerging.pdf. Accessed on 26 Feb 2018

Wright R (1983) Acidification of freshwater in Europe. Water Qual Bull 8:137–142

Xu F-L, Jorgensen SE, Shimizu Y, Silow E (2013) Persistent organic pollutants in fresh water ecosystems. Sci World J (303815), 2 pp

Yeomans J, Bowater D (2016) One year on, Brazil battles to rebuild after the Samarco mining disaster. Available via http://www.telegraph.co.uk/business/2016/10/15/one-year-on-brazil-battles-to-rebuild-after-the-samarco-mining-d/. Accessed on 26 Feb 2018

Zielinski S (2015) Pink salmon threatened by freshwater acidification. Available via https://www.sciencenews.org/blog/wild-things/pink-salmon-threatened-freshwater-acidification. Accessed on 26 Feb 2018

Chapter 6
Water as a Source of Conflict and Global Risk

It is clear that water has two fundamental functions namely being a prerequisite for life on Earth as well as being an economic resource or commodity for further development. These two roles are often in conflict all around the world mainly due to continued competition between different water usages, human livelihoods as well as the environment. The continued competition between water usages has consequently led to the exploitation of water through different human activities which has in turn increased the risk and placed great pressure specifically on aquatic ecosystems and the life which they support (Pimentel et al. 2010). Conflicts may therefore arise on various levels between agriculture, industrial and domestic water use sectors as well as the natural reserve due to increased water stress. Increased water stress may be caused by natural events such as droughts or by human activities which may include but not limited to overuse of surface or groundwater as well as water pollution which lessens the amount of usable water for different water uses and creates additional vulnerabilities and risks.

Different types of subnational as well as national conflicts may arise all over the world from water scarcity and be accompanied with immense environmental degradation, socio-economic consequences and risks. The chapter will evaluate water as a source of conflict with the use of current case studies. The chapter will also focus and conclude part one of the book by evaluating water as a global risk. Focus will be placed on the cost of current and predicted water-related problems such as water stress and water-related diseases which may pose significant risks for future environmental sustainability, environmental and human health as well as socio-economic growth.

6.1 Water as a Source of Conflict

The US Central Intelligence Agency issued a warning decades ago that many countries will experience water problems/shortages, poor water quality or floods which will risk instability, state failure and increase regional tensions. In 2001 the then UN

Secretary-General, Kofi Annan highlighted with concern that the fierce competition for freshwater resources may become a source of conflict and wars in future, however, he revised his statement a year later saying that water can rather be a "catalyst for cooperation". After decades of "water wars" threats which never materialised, these conflicting statements emphasise the complexity of how water and conflict interact (Swanson 2014).

Some scholars highlight the ancient Babylonian conflict which occurred 4,500 years ago being the only true "water war" which has ever occurred. The fact that one-fifth of the world's population face water scarcity and another 1.6 billion people live in countries whose infrastructure is too weak to get water where it is needed, it is surprising that major conflicts have not occurred.

Water as a prerequisite for life on Earth as well as being an economic resource or commodity for further development fulfils many roles in our environment and society. Water is a fundamental resource for drinking, food production, fisheries and transport, sanitation, it is a solvent and cooling agent and lastly has religious significance. Therefore, when you think of water and the management thereof, you also need to consider its role in various sectors as you are managing all of these factors or sectors as well. The following sections will evaluate the role of water stress in creating conflict, the relationships which exist between water stress and civil unrest or military conflict, how competing interests may contribute to water stress and ultimately tensions and lastly the role which climate change might play in creating future possible conflicts or tensions with the use of detailed case studies.

6.2 Role of Increased Water Stress in Creating Conflict

In 2012 an unclassified version of a USA National Intelligence Council report on Global Water Security was released which stated that, without more effective water resources management, between now and 2040, worldwide fresh water availability will not meet the demand. This report further indicated that "while wars over water are unlikely within the next 10 years, water challenges—shortages, poor water quality, floods—will likely increase the risk of instability and state failure, exacerbate regional tensions, and distract countries from working with the USA on important policy objectives" and that "water problems will hinder the ability of key countries to produce food and generate energy, posing a risk to global food markets and hobbling economic growth". The concern related to the possible effects of global water shortages on political stability is therefore not new and have been receiving attention for the past three decades around the world (Intelligence Community Assessment 2012).

Approximately half of the world's population, as well as a large number of ecosystems around the world, are affected by water problems which in turn create stresses influencing the stability of the affected communities and have the potential to exacerbate festering antagonisms and quarrels. Rapid urban developments which have resulted in numerous informal settlements, especially in the case of developing countries, have been accompanied by lack of domestic waste disposal, sanitation and sewage effluent systems and ultimately forces people to inhabit areas which have very limited sanitation and water supply. The world's poor population, particularly children, are especially negatively affected by unhygienic insufficient water. The situation is further exacerbated in communities where there exists a competition for adequate clean water supply and consequently leads to further public and private discord. Lack of clean and reliable water also ultimately affects food security as it is intimately linked with water. Serious problems which arise from inadequate water can last for generations and these externally imposed stresses such as these can ultimately lead to social unrest, political instability and in some cases even armed conflict.

Droughts or periods of drier than normal conditions also lead to water-related problems and increased water stress. In recent years multiple regions around the world have experienced droughts which in some cases have led to protests or conflicts within communities themselves and/or with other competing water sectors. The following case studies will look at the role of drought or declining water resources in creating different possible conflicts in both developed and developing countries.

6.2.1 Parched Western USA Region: Severe Drought, Water Rights and Societal Impacts

Developed countries such as the USA will be unable to avoid water supply problems and future instability due to consequences associated with climate change as well as the occurrence of more frequent extreme weather events. The western region of the USA is a leading example of how continued drought can attribute to various types of instabilities or possible conflicts within the region.

The drought has been described as the worst in history, has affected the local and national economy since 2012 and have cost various states millions of dollars. The persistent drought has also been accompanied with a series of legal and political battles over who controls the now precious water resources. Farmers which pumped water from the desiccated Brazos River were ordered to shut down their pumps due to the sprawling photochemical complex having more senior rights. This was implemented by the government to make up the deficit and gave cities and power plants along the river a pass on the basis of public health and safety overriding farmers own water rights. The farmers in the region have consequently taken the government to court over the issue and have been winning.

The population of the arid Western region has always argued over its water resources however the persistent drought has intensified these struggles and the continued growth and thirst of these western urban areas have raised the stakes to new levels. Some of these struggles have led to legal actions or stricter restrictions and have included the following:

- The state has cut off deliveries of river water to rice farmers along the Gulf coast southwest of Houston for 3 years to sustain reservoirs that supply the ever-expanding Austin urban region.
- Lawsuits have also been issued in Nevada to block a pipeline which would supply Las Vegas with groundwater from the aquifer straddling the Nevada-Utah border.
- Strict restrictions are being imposed in Colorado on requests to ship water across the Rocky Mountains to Denver and the rest of the state due to fear of their existence. Local water rights sales have consequently been blocked to Denver's fast-growing suburbs.
- Activists in Arizona are attempting to stop plans related to the pumping of groundwater used by a vast housing development which would reduce the water levels of a protected river. Kansas has accused Colorado and Nebraska of allowing farmers to divert Kansas' share of the Republican River. There is a similar dispute between Mexico and Texas (Wines 2014).

The Californian region has received most media attention due to the drought it has experienced the last couple of years. The drought has been attributed to unusually low snowfall which occurred across the state mainly due to increasing winter temperatures over the recent years and cost California US$2.7 billion in 2015. Drier conditions have also led to larger and more frequent fires posing danger to people and property as well as other hazards such as landslides and floods (Poppick 2014; Aleem 2015; Kasler and Reese 2015). The drought has also led to the deposition of soil and heavy metals into the already strained water resources, increased turbidity and may have further financial implications in terms of increasing the water treatment costs. The agricultural sector has been the hardest hit as it is the biggest water user (80%) in the region. Farmers stand to lose US$810 million from keeping fields fallow, a further US$453 million on pumping groundwater and likely to lose 17,000 agricultural jobs due to the drought (Aleem 2015; Kasler and Reese 2015).

Societal impacts have included the loss of jobs especially within the agricultural sector and a rise in food prices. Political consequences have included that the government has had to request urban consumers to decrease water usage by 25%. Some inequality of access to water resources has arisen where 48% of wealthy homeowners with income above $100,000 have stated that it would be too difficult to conserve water. This has consequently led to heated debates within society whereby rich people do not mind paying a $100 fine for consuming too much and ordinary people are blamed for consuming too much. An increase in class inequality as well as differing perspectives on water consumption has occurred which has placed immense pressure on the government to manage this precarious situation with a state historically characterised by injustice and racism with the allocation of water during the Gold Rush eras in the region (Aleem 2015; Johnson 2015; Kasler and Reese 2015).

The main conflict is between farming and fishing industries and has been exacerbated by the drought causing the possible extinction of native fish species such as the delta smelt as well as threatening other native fish such as the longfin smelt, green sturgeon and winter-run Chinook salmon. Delta smelt populations have been declining for decades due to invasive predators, pollution, habitat loss and increased water exports to farms and cities. The drought has worsened conditions by reducing freshwater flows and raising water temperatures and led to government regularly cutting water exports from the delta to protect the fish species and other threatened fish from being sucked into the giant pumps that send water south. Farmers are of the opinion that too much water has been wasted on these fish and scientists and environmentalists state that it needs protection due to it being an important indicator species for the delta's health. Some almond farmers have had to pull out some of their almond trees due to them not being able to obtain enough water for irrigation (Kahn 2015; Pedroncelli 2015).

The additional conflict between farmers and the urban population has been centred around government ordering urban areas to cut their water use by 25%. Urban residents have consequently developed the perception that the agricultural sector is getting off easier and that this sector should be subject to more regulations. However, it should be noted that due to the drought, farms have been allocated 0% share of the water from irrigation canals of the Central Valley Project and have led that half a million acres of farmland have had to be laid fallow in 2014 (Walker 2015).

Main critiques have been aimed at almond farmers who have continued to plant almond and other nut trees which require annual watering. California grows 80% of the world's almonds and is by far the biggest exporter of processed fruit and nuts. State officials have had to defend the agricultural industry and invoked globalisation for the root cause as more than two-thirds of almond crop is exported, much of it to China (Walker 2015). This has also raised the debate of water efficiency in different farming sectors especially between almonds and beef or dairy production whereby it has been emphasised that it takes more than 380 L of water to produce a 28 g of beef, compared with less than 190 L for 28 g of almonds. The decrease in water use in the almond farming sector is mainly due to advances in irrigation technology which have lowered their water demand by a third. The conflict has extended to that opprobrium ought instead to be heaped on alfalfa hay, a low-value crop sold as feed to dairies overseas, which takes up even more of the state's agricultural water supply than do almonds, 15% in recent years. Almond farmers further defend their stance by suggesting that almonds generate more jobs for the state economy per unit of water consumed than alfalfa, rice, beans or corn (Kahn 2015; Walker 2015).

Different types conflicts have therefore risen from the persistent drought and these conflicts and debates will continue within the region and increase in intensity with the continuation of the drought as well as the persistent and increased impacts of water stress, changes in climate and the insatiable societal demands while trying to ensure a healthy environment. A positive outcome is that the drought has forced farmers to adopt more efficient water management technologies and practices that helped boost the revenue within the limited available water and urban consumers have also been requested to cut their water use down by 25% by the government (Cooley et al. 2015; Johnson 2015). The adaptations made have consequently buffered the region's economy and job levels. Some of these adaptations and responses will build resilience, while others will have lasting and damaging consequences to the region's population, its ecosystems and future generations.

6.2.2 Possible Future Water Conflicts in India: Persistent Drought and Continued Poor Water Management

India has been facing one of its most serious droughts and it has been estimated that around 330 million people are likely to be affected by acute water shortages. It has consequently placed the country's available water resources high on the public agenda. Four of the ten drought-hit states within the country face chronic conflicts due to poor water management and the diversion of water to urban areas. Even though the country receives considerable rainfall most years through the annual monsoon, the rainfall falls in particular areas, for a short period of time. This consequently leads to distress in terms of flooding due to the drought (Vira 2016).

The drought has forced the Indian government to resurrect plans to attempt and link major river basins through the Interlinking of Rivers project. Critics have however suggested that it will most probably be unsuccessful and lead to ecological and social disruptions. The continued inadequate rainfall in the country has led to the drying up of reservoirs and village water bodies especially in the grain-growing regions of southern India. The country is experiencing the worst drought in 140 years and has led to people having to leave villages and move to urban areas, people not having food or water and no fodder for livestock. The Cauvery River, which was once an 800-km river on which millions of farmers depend, has become dust tracts in several sections. This has also led to soil erosion and further reduction in rainfall. The removal of the natural storage of monsoon rainfall has consequently led to floods and further drought. This in combination with the over-extraction beyond the river's capacity has left the river dry and placed numerous people's livelihoods in jeopardy (Vira 2016).

The assumption of having larger cities by diverting water from hundreds of miles away may kill all rivers in the country if all rivers are diverted to urban and industrial areas. Other effects of the damming of rivers may include coastal erosion, deforestation as well as the displacement of already vulnerable people and exacerbate possible impacts of climate change. Centralised irrigation systems and large dams which have been introduced have led to immense soil erosion and over-extraction of underground aquifers have depleted the water table. Bauxite mining has also contributed to the collapse of groundwater levels and has left hills bare and arid.

This water crisis has led to the failure of crops and has caused desperate farmers to obtain loans with exorbitant interest rates to purchase food, seeds, fertiliser and equipment. Drought-hit farmers from Tamil Nadu consequently protested for farm loan waivers but few state governments have conceded. Furthermore, upstream states like Karnataka have refused to share the Cauvery River's water with neighbouring or downstream villages which will cause crops to fail and could transform them into deserts. Violence in the streets broke out after Karnataka decided not to comply with the supreme court ruling of releasing more water. The drying up of the Cauvery River could be India's greatest human catastrophe ever if these issues are not resolved and local water preservation and community-driven water management systems are not considered and implemented (Ng and Mukherjee 2017).

The management of water demands has not been prioritised within the country and water-thirsty crops still dominate the dry regions of the country. Farmers receive energy subsidies which allow them to over-pump already depleted aquifers. Distributional equity issues are also prominent within the country where the poor in urban contexts pay more per litre for erratic and unreliable water while richer neighbours have the luxury of definite water supply.

The water conflict possibility lies within the context of the transboundary nature of the water issues within the Hindu Kush Himalayan region which spans across eight countries, supports ten major river systems and potentially affecting 1.5 billion people. Transboundary cooperation is therefore vital to manage fragile resources which are further threatened by uncertain impacts of climate change. Despite three major wars since India's independence, India and Pakistan have managed to maintain some cooperation however it has been suggested that regional conflict over water resources will worsen. The possibility of worsened regional conflicts is much dependent on China as it is the dominant upstream water controller in the region. India will, therefore, have to focus more attention on its already fragile water resources as these issues span over social and economic life. An integrated water management approach will be required to address sustainability, land use management, agricultural strategies, improved demand management as well as distribution and pricing of water especially with the growing pressures accompanied with climate change and the constant migration and population growth (Vira 2016).

6.3 Water Stress and Civil Unrest or Military Conflict

Water resources have rarely been the sole source of violent conflict or war, however, there are complex and real links between water and conflict. There is a long history of tensions and violence over access to water resources, attacks on water systems as well as the use of water systems as weapons during war. Water has played different roles throughout human history in terms of creating unrest and in terms of military conflict. During major global conflicts, clean water supplies served as direct military targets or military tools. In modern times which have lacked all-out global war, regional or local battles for economic and social development has dominated along with terrorist activities which have centred around controlling local water supplies to promote their ideological religious or ethnic factions.

While future large-scale wars over water are not anticipated, water scarcity has been deemed as a factor which can increase regional conflicts and tensions, encourage border disputes and possibly be a focus of terrorism, local tribal and ethnic warfare as well as political disputes in terms of competing economic developments. Water disputes which have occurred in the past decade have not produced large-scale global war, however, regional conflicts and local wars have often used water as part of a ploy to advance political goals. The following case studies will look at the role of increased water stress in combination with civil unrest or military conflicts.

6.3.1 East Africa Water Wars and Prolonged Civil Unrest in Sudan, Darfur

The East African region has been plagued by rising temperatures, droughts and increased water scarcity. This in combination with growing populations especially where rivers are shared by more than one country, constant fight for water usage rights and continued water shortages have led to multiple examples of water conflicts in the region. Water has consequently been used as a weapon to obtain either political or socio-economic goals.

The drought between 2004 and 2006 over the East African region affected approximately 11 million people, killed large numbers of livestock in the region and forced the Kenyan and Ethiopian governments to intervene in numerous skirmishes over water in their countries. Military forces and police needed to intervene in some cases to pacify battles specifically around wells. Significant fighting over groundwater resources occurred within Ethiopia during this period. Conflict developed between two clans and gave rise to "well warlords" and "well warriors". The extensive violence was labelled as the "war of the well" and led to the death of 250 people and many injured.

The history of Sudan especially the Darfur region has been characterised by civil war, famine, coups as well as tyranny. Years of civil unrest in the Darfur region has been dominated by the intentional bombing of wells around villages such as Tina and contaminated in Khasan Basao in 2003 and 2004. The previous stance for the main cause for the prolonged conflict was solely placed civil unrest, on the Muslim government in the north which has been engaging in civil war with rebels in the Christian or animist south as well as genocide against ethnic groups. This view has however changed in recent times with the sharing of new data as well as increased focus on declining water resources and arable land within the region (Muhammad 2010; Schlein 2011).

The region has faced immense ecological crises which predominantly include water scarcity and desertification, displacing rural populations through changing landscapes and lack of agricultural production. The livelihoods of the country's population are dependent on agricultural production as the sector accounts for 97% of water use. Farming practices have however degraded the environment, reduced arable soil and have caused desertification to spread. The possible causes for the continued conflict in the Darfur region have been attributed to the following:

- Spread of deserts southwards by an average of 100 km over the past four decades;
- Overgrazing of fragile soils causing widespread land degradation;
- The "deforestation crisis" which has led to 12% loss of the country's forest cover in just over 15 years;
- Declining and highly irregular patterns of rainfall particularly over the Darfur and Kordofan states. The rainfall in the Northern Darfur region has decreased by a third over the last 80 years (Worldwatch Institute 2018).

6.3 Water Stress and Civil Unrest or Military Conflict

The conflict over arable land and declining water resources have consequently contributed to instability within the region. It has been linked to a breakdown of law and order, associated flow of weaponry and the overflow of conflict from neighbouring countries, mainly Chad, and ultimately continued civil unrest within the region. In some cases, water resources have been specifically targeted and ultimately used as a weapon to obtain either political or socio-economic goals. The prolonged drought of 1983, the famine of 1984–1985 and the creation of severe water scarcity in the Darfur region has however been labelled as the root cause for the conflict in the region. Farmers and Arabic nomads have long been competing for limited water resources and grazing land due to the ever-expanding Sahara Desert which has also contributed to ongoing conflict (Polgreen 2007).

Water resources have therefore not always been the main target but rather the main contributing cause for conflict in the region. The perceived misallocation and unavailability of water in the region have instigated clashes which have been labelled as development disputes.

Due to water being labelled as being the main root cause of these conflicts, a need has risen to address this issue as people believe that water can consequently be used as an instrument for peace. The implementation of well managed and equitably distributed water resources has been given as possible instrument for sustainable peace within the region. These steps can however not be taken in isolation.

It needs to be highlighted that the conflicts experienced within the region is multifaceted and cannot be solved by just addressing current water stress. Civil unrest as well as instability, increased desertification, the associated decrease of arable land which threatens most of the region's population's livelihoods as well as the future effects of climate change on the region also needs to be considered and included as these factors also play a significant role which can not be ignored.

6.3.2 *Syria: A Country Unravelled*

Syria is one of the driest countries in the world receiving less than 250 mm rainfall annually and is a water scarce country. Pressures on the country's water resources have been increasing over the past couple of decades. Approximately 60% of its renewable surface and groundwater resources originate outside of Syria's borders (Frenken 2009) and all of its major rivers are shared with neighbouring countries. Tensions have occurred between Jordan and Syria since 1990 over the construction of Syrian dams. Turkey and Syria also have long-standing disputes over the management of the Euphrates River and these tensions have worsened over the past decades with the completion of the Ataturk Dam and a decrease in rainfall in the region. Population dynamics have also played a major role as the population has increased from 3 million in 1950s to 22 million 2012 which has decreased the country's total per capita renewable water availability to under 760 m^3 to a level categorised as scarce (Gleick 2014).

In addition to having little overall freshwater in proportion to demands, the region also experiences high natural hydrologic variability. Syria has experienced six significant droughts between 1900 and 2005 where five lasted for only one season and the sixth lasted two seasons (Mohtadi 2013). Syria experienced a multi-season, multi-year period of extreme drought from 2006 to 2011 which led to agricultural failures, economic dislocations as well as population displacement. The current civil war has been attributed to the drought including agricultural failures, water shortages as well as water mismanagement which contributed to deterioration of social structures and spurring violence (FAO 2012; Femia and Werrell 2013; Mhanna 2013).

The persistent and severe drought has been combined with multiyear crop failures and related economic deterioration has led to significant dislocation and migration of rural communities to urban areas which further contributed to urban unemployment as well as economic dislocations and social unrest. These impacts have been described as the "perfect storm" when combined with other economic and social pressures. The return of the drought in 2011 worsened the situation even further and drove millions of people into food insecurity and forced more than 1.5 million people (mostly agricultural workers and family farmers) to migrate from rural land to cities and camps on the outskirts of the country's major cities. Poverty and food insecurity still increased and conditions worsened due to poor water management decisions, poor planning and policy errors. Most of Syria's irrigation still relies on highly inefficient flood irrigation which still needs to be modernised and half of all irrigation is dependent on groundwater which is also overpumped (78% of groundwater withdrawals are unsustainable) and have led to the dropping of groundwater levels and increased production costs. These additional factors to the drought added to further economic and political uncertainty (Gleick 2014).

The failure to implement economic measures to combat or address the effects of the drought as well as the associated economic and environmental conditions were all drivers for massive mobilisations of dissent and subsequent political unrest (Saleeby 2012). The extensive exploitation of groundwater has also contributed to the problem as it has led to substantial drops in water levels and in some cases contamination by salts and nitrates, making wells unfit for human use.

The development of unrest in Syria was accompanied by worsened violence. In 2011, Syria experienced disruption in a wave of political unrest over North Africa and the Middle East due to dramatic changes in the availability and cost of food (Arab Spring). It was dominated by religious differences, failure of the ruling regime to address increasing unemployment and social injustice contributed to further social unrest. Impacts on urban water distribution systems were reported with intentional attacks on water systems due to their strategic value. Many of the reservoirs have been intensely fought over and reservoirs under rebel control are badly managed as they lack proper skills or staff to operate them properly. The civil unrest has created millions of people to become refugees. The deterioration of water security in Syria may cause increasing disease and fuel migration which may consequently deepen pollution and water scarcity in neighbouring countries such as Jordan (Jones 2017).

6.3.3 Yemen: Humanitarian Crisis and a Non-existent State

Yemen is located in the Arabian Peninsula and is the poorest country in the Middle East. The southern region is characterised by expanding deserts, being the driest and have been plagued by persistent conflict since decolonisation. The country's human population has grown exponentially since 1980 from 8 million to over 27 million people and consequently placed great strain on the already stressed water resources. Approximately 50% of the country's population struggle to access or purchase enough clean water for drinking and food production. Consequently, 14.7 million people in the country depend on humanitarian aid (Cruickshank 2013; Whitehead 2015).

The change to a global trade/cash economy together with farmers changing to cash crops which use large amounts of water and the production of a stimulant drug called "qat" at the cost for food for the average citizen have also increased water use. New drilling techniques were introduced to try and tap into its fossil water reserves and led to unsustainable extraction from these aquifers, drills having to go deeper and deeper and water prices increasing. The abandonment of traditional irrigation methods to favouring the pumping of groundwater has caused aquifers to become depleted at a very fast rate. Yemen has always been a water-stressed country, however, the vast expansion in its human population and poor water management has exacerbated the problem (Cruickshank 2013; Whitehead 2015).

The country has further also been plagued by persistent drought which is being exacerbated by climate change. Water and food have consequently become scarce and have led to the displacement of populations and violence. Some experts have warned that Yemen might be the first modern country to run out of usable water and that it can occur within the next decade.

Tensions over strained water resources have reached extreme levels and led to the fracture of Yemen along sectarian and regional lines. Protests over rising petrol prices and rising prices of water transported by trucks started in 2014 by the Houthi minority and led to a revolution against the sitting government which fled south and formed an Arab Coalition with Saudi Arabia. The country has been in the midst of a major civil war after the Arab Spring uprisings in 2012. A Saudi-led coalition has started an aerial campaign and will continue air strikes till they have achieved their objective of saving the ousted president Abd-Rabbu Mansour Hadi's government (Whitehead 2015; Lopour 2016, 2017).

The civil war has exacerbated the country's water problems and has placed more people in needing humanitarian aid than any other country in the world. Water has become a weapon of war as millions of people rely on humanitarian aid to meet basic water needs and fuel shortages have led to people relying on water truck deliveries as they are unable to pump water. There have been reports of children being bombed or shot while waiting in line for water cans. The civil unrest has also delayed and restricted trucks carrying humanitarian aid through naval blockades administered by the Arab Coalition. A substantial portion of the country's infrastructure has been destroyed. Key strategic dams, viaducts and water treatment plants have been destroyed by the persistent conflict initiating unprecedented water scarcity levels (Whitehead 2015; Lopour 2016, 2017).

The country's health system has also collapsed and unsafe drinking water has put its population at high risk for communicable diseases such as cholera and dysentery. The country has consequently been experiencing an outbreak of cholera since October 2016 as many hospitals and clinics have been destroyed or shuttered and its damaged critical water, sanitation and hygiene infrastructure has been exacerbating the problem.

Even though warring political factions may reach a political solution in the near term, the underlying factors of the country's water crisis is not set to improve due to a weak to non-existent state. The lack of governance and mismanagement will restrict necessary maintenance of infrastructure and conflict resolutions. Additionally, the impacts of climate change will also intensify water problems in the coming years as the warmer climate and increased evaporation will reduce the overall water reaching the country's rivers and aquifers. Immense assistance will be required to try and meet the population's basic needs as the country is continued to be withered by war, water shortages, government mismanagement and climate change. Solutions which have been suggested to be implemented once political stability has been achieved include the treatment of shallow coastal wells to make these drinkable and safe and starting with rainwater harvesting in mountain areas as a starting point however the government will have a major job on their hands to try and address the country's immense water problems (Whitehead 2015; Lopour 2016, 2017).

The humanitarian crisis has been described as worse than Syria and has not received as much attention. Several other countries in the Middle East region are also experiencing water shortages making the issues within Yemen critical for regional stability.

6.4 Water Stress and Competing Interests

Water stress combined with competing interests for limited resources can lead to political turmoil on a regional, national and in some cases even international levels. Conflict may arise when water resources such as rivers are used as a political instrument or as potential threat. There are numerous regional cases where rivers flow through several adjacent nations. The strengths, weaknesses and absences of existing treaties between these political entities may in some cases create tensions. Three case studies will now be discussed as examples of disputes or possible transboundary conflicts over surface water.

6.4.1 The Brahmaputra River: Sinking of China and India Relations

The Brahmaputra River originates in Tibet and flows through India before merging with the Ganges and draining into the Bay of Bengal in Bangladesh. The river is an important water resource for China as it provides hydroelectricity and it is a key agricultural lifeline for India and Bangladesh within a region characterised by overpopulation and aridity. It is mainly important for the agricultural industry in India's Assam Plains and concerns have been raised regarding multiple hydrological plants which China is in various stages of construction on its Tibetan plateau as it is believed by some that these projects may reduce flow of the river in India which can compound the already fragile water situation in the affected areas (Ramachandran 2015).

Potential conflict in the southern Asia region is particularly focussed on the borders of India, Pakistan and China. Tensions are rising over rivers which run cross-border such as the Indus River (flowing from India to Pakistan) and the Tsango/Brahmaputra River system (flowing from China to India). In terms of India and Pakistan, an Indus Water Treaty was developed in 1960 when Pakistan was concerned over existing and planned Indian dams which can potentially limit water in critical growing seasons. In 2011 an editorial in a Pakistan Newspaper Nawa-i-Waqt stated "Pakistan should convey to India that a war is possible on the issue of water and this time the war will be a nuclear one". The statement is primarily based on the fear that India can "switch off" the Indus to make Pakistan solely dependent on India which is considered as a possible "water bomb" (Ramachandran 2015).

On the other side of the spectrum, India claims that China will divert the Brahmaputra River and place millions of livelihoods in jeopardy. Recent years have seen the two governments deciding to tame the river by building hydropower dams in an effort to try and support their ever-growing energy requirements and over 160 memoranda of understanding have been signed between the Arunachal Pradesh state government and private and public dam-building companies to build medium and large dams in the state. The size and scale of these proposed developments are unprecedented in the history of northeast India and indigenous communities within the affected region are concerned that they will face major social and economic upheaval with the approval of all these projects. India's policy of pursuing hydropower development in the region could also have negative impacts on bilateral relations (Patranobis 2016).

China's plans to divert the river have also set off anxiety in the lower riparian states of India and Bangladesh as it will have repercussions for water flow but also agriculture, ecology, lives and livelihoods downstream and undermine Sino–Indian relations. China maintains that all their proposed hydropower projects are run-of-the-river projects which involve no storage or diversion, however, others are less optimistic and still consider these projects of utmost concern as they will influence lower riparian countries negatively. India's damming of the Ganges River which has reduced river flow to Bangladesh has also contributed to tensions (Lumaye et al. 2016).

China is also planning the northward rerouting of the river's waters at the Great Bend which can result in a significant decrease in the river's water level as it enters India and has serious impacts on agriculture and fishing in downstream areas and increase salinity. However, others are still of the opinion that the impact of the diversion will not be severe. Analysts in the region have warned that "water wars" can break out between India and China as upstream dams, barrages, canals and irrigation systems may fashion water into a political weapon or subtly in peacement to show dissatisfaction with a co-riparian state (Ramachandran 2015).

China has however announced in late 2016 that it will block a tributary (Xaibuqu River) of the river to construct one of the country's most expensive hydroelectric projects. The impact, if any, of the blocking is not immediately clear. The Xaibuqu River is not a trans-border one and does not fall under the bilateral mechanism between China and India (Ramachandran 2015).

China's proposed dams are key to national economic and energy development priorities and have placed less emphasis on international ramifications of planned diversions. Even though there is no comprehensive bilateral treaty in place for the sustainable management of the river, an information sharing agreement for hydrological data has at least been agreed upon between the two governments. This will hopefully improve communication on the issue and lessen suspicion and tension between the riparian states. Cooperation has therefore been expanded between Beijing and New Delhi with this agreement of sharing hydrological/flood data during the flood season after a decade of negotiations. India has officially accepted that China's dams do not pose danger to downstream water flows and have consequently created a nexus between hydrological and geopolitical worries. However, the Brahmaputra River will remain a potential source of friction between these preeminent rising powers as India remains sensitive to its control and development of its restive northeast region through which the river runs. Multilateral cooperation between the riparian states, therefore, needs to become entrenched to ensure that the region does not move slowly closer to "water wars" (Ramachandran 2015; Wuthnow 2016).

6.4.2 The Renaissance Dam and Nile River: Ethiopia and Egypt

The Ethiopian government announced in 2011 that they are planning to build the "Grand Ethiopian Renaissance Dam". The dam will predominantly be used for hydroelectricity, have a capacity of 6,000 MW and cost approximately $4.1 billion. The hydroelectric dam will be constructed on the Blue Nile near the border with Sudan and is meant to capitalise on the country's considerable hydroelectric potential as well as provide electricity for themselves and surrounding regional populations (Michel 2013).

The construction of the dam has however angered Egyptian authorities who claim that it may limit their water supply and decrease their farmland by approximately

25%. There are also fears that Ethiopia will trade one problem for another as it may jeopardise its own water security by increasing the volatility of a river which has a history of being difficult to predict.

The potential impacts of the construction of the Renaissance Dam will particularly be on downstream water supplies which is a grave concern for Egypt who has consistently opposed the construction since it was proposed. Egypt's legal argument concedes to treaties from 1929 and 1959 which guarantee it two-thirds of the Nile's water along with veto rights on any upstream projects. Ethiopia ignored this right when they unilaterally proceeded with the construction. Ethiopia has partially diverted the course of the Blue Nile and initial filling has started in 2017 with the completion of construction (Michel 2013; Whittington 2016).

The building of the dam raises existential alarms for Egypt as the country receives almost no rainfall and depends on it for 97% of its renewable water resources. The Nile also depends on Ethiopia as more than four-fifths of the river's water falls as rain in the Ethiopian highlands.

The development of a multilateral approach to develop the Nile has failed thus far. The failure was especially evident in the 2010 Cooperative Framework Agreement where upriver countries joined together against downriver countries which refuse to give up their historical rights even though economic power and dynamics have changed in the region. Ethiopia maintains that the Renaissance Dam will not have a significant impact on its neighbouring countries however Egypt fears that it might diminish its vital water supplies. Despite constant disputes it has been stated that there is little chance that the two countries might clash swords over the Nile as both stand to lose too much from war. Even though the chances of conflict might be low, the jarring demands on the Nile's shared waters are real and symptomatic of other similar conflicting claims over scarce water resources in other regions as well (Whittington 2016).

6.4.3 *Ilisu Dam and the Tigris River: The Case of Turkey and Iraq*

Water has started to play a vital role and has been described as the main cause of disputes between Middle Eastern countries in recent years as evident in some of the previous case studies. Multiple years of war, thoughtless water supply management, unchecked population growth, misguided agricultural policies as well as the supporting of consumption have led to a growing water crisis in the region. Most of these countries are also comparatively downstream riparian countries which lack water resources and has led to a natural convergence of states which like to hold power and authority on limited available water resources in the region (Bari 2016). The case of Turkey and Iraq is an example of these growing conflicts within the region.

The headwaters of the Tigris and Euphrates Rivers begin in Turkey and Syria and flow into Iraq. The control of these rivers is dependent on the release of water from upstream dams such as the Atatürk Dam which is the centrepiece of 22 Turkish dams. A concern has been raised recently that these dams in Turkey have not conformed to "international guidelines designed to prevent human rights violations through development and infrastructure projects" and a UN report notes with alarm that "the Turkish government has performed no assessment of the environmental and social impacts of these dams, perhaps because they would mostly impinge on already marginalised groups such as the rural poor, nomads, the Alevi, and the Kurds in violation of Article 2.2 of the International Covenant on Economic, Social and Cultural Rights (United Nations High Commissioner for Human Rights 1966)".

This is a major concern for downstream riparian countries but more specifically for Iraq which is facing significant threat of water shortages attributed to internal and external challenges. These challenges include poor water management, internal political conflicts as well as unstable relationships with neighbouring countries which include Turkey, Iran and Syria. Increased water shortages can severely impair Iraq's economy and may pose unforeseen environmental issues. Water has become a key factor in the creation of either peace or war within the region. Iraq, therefore, needs to prioritise the development of strict hydro-policies to try and mitigate increasing risks (Al-Muqdadi et al. 2016).

The Erdogan government has been keen to approve the final part of the South eastern Anatolian Project which includes the Ilisu Dam on the Tigris River located near the Syrian border. The Ilisu Dam is the most recent of many Turkish projects aimed at obtaining the hydroelectric potential of both the Tigris and Euphrates rivers. The Ilisu Dam will generate 1,200 MW or approximately 2% of Turkey's energy needs (Bari 2016).

Iraq will be the most affected by Turkey's upstream activities and Syria to a lesser extent. Iraq has always historically relished on most of the share of these rivers which have supplied seasonal marshlands which are used to grow food. These waters have however receded over the past decade even before the completion of the Ilisu Dam. Currently, northern Iraq and Syria are experiencing extended droughts and some analysts have mentioned that this could possibly have contributed to the rise of Islamic State of Iraq and Syria or ISIS in the region. Some extreme projections have shown that the combination of climate change as well as upstream dam activity in the region will lead to the Tigris and Euphrates rivers not having enough water to reach the ocean as early as 2040. This has consequently further raised concerns and uncertainty within Iraq regarding its future water resources (Al-Muqdadi et al. 2016; Bari 2016).

Iraq has accused Turkey of having a hidden hydro-political agenda and Turkey complain that Iraq's claims are legally unfounded due to these rivers originating on Turkish soil. These two countries have failed to reach a consensus, Turkey has continued with its plans while Iraq has lost time as well as resources.

Assumptions have been made that Turkey is aiming to accomplish two goals at once. These goals include first to gain control of water courses which belong to Syria and Iraq and second to restrict water shares of these downstream countries to such an extent that they are forced to depend on them politically. Turkey will gain long-term advantages and could assist them in overcoming energy shortages and ultimately

6.4 Water Stress and Competing Interests

establish them as a key player in Middle East politics. Turkey has justified their actions through the following arguments:

- Downstream countries do not have proper water management of their water resources and most freshwater is wasted. These countries will benefit from the project as it will prevent floods as well as water waste.
- Turkey has the right to control rivers' flow as 90% of Euphrates' and 50% of Tigris' total annual flow originates on Turkish soil.
- The UN Convention does not have binding legal status and additionally, all natural resources should be shared if the convention was to be followed literally.
- Iraq should have a lower share of water as irrigable land within the Euphrates basin does not exceed 1.95 million hectares. Moreover, most of this river basin is infertile and it would be useless to expend more water for them.
- Turkey has proposed that water from the Tigris and Euphrates Rivers should be allocated for the purpose of alleviating water shortages but this was rejected by Iraq.
- Lastly, Turkey uses these dams for community development, power generation as well as the management of demographic changes within the country and not for hydro-political gain. The dams would therefore not pose a potential threat to these downstream countries (Al-Muqdadi et al. 2016).

The Economic Commission for Europe's (ECE) Convention on the Protection and Use of Transboundary Watercourses and International Lakes entered into force in 2013 and Iraq signed this convention in June 2015. The signing of the convention may strengthen the legality of Iraq's claims and give a possibility of a win–win position. This will, however, require the establishment of a healthy negotiating environment and a change in the prevailing discourse from a political to a scientific level. Turkey has also shown legal concerns over water resource management and Iraq has not adequately managed its water resources. The arguments given above by Turkey would appear valid under the assumption that it does not operate with political motives however many indicators support the view that Turkey is acting with hydro-political motives which have led to conflicts between the two countries (Al-Masri 2014; Al-Muqdadi et al. 2016).

The dam construction site was militarised in December 2014 which led to further political tensions and human right violation despite the predicted dramatic social, cultural and ecological impacts in the affected downstream regions. The dam construction site was militarised after construction halted from August 2014 due to the resignation of workers. Workers resigned due to the kidnapping of subcontractors by guerrilla forces of the People's Defending Forces. Approximately 80% of construction was completed at this time but the hydroelectric power plant had not been constructed. Turkish people indicated that they did not want to work on the Ilisu Dam construction site as it is a threat to their lives and subcontractors consequently employed workers mainly from non-Kurdish provinces within the Republic of Turkey (Al-Masri 2014; Ayboga 2015).

Syria and Iraq already went to war with Turkey twice during the period of 1975–1991 with military action and have gone to war once over waters of Tigris and Euphrates Rivers and it seems like this might occur again. The current wars in Syria and Iraq have intensified the situation as water networks and infrastructure have been targeted and in some cases destroyed. Some of these affected water networks have also now come under the control of ISIS and Kurdish separatists have threatened Turkey by stating that they will attack its dams as a means of retaliation over clashes (Ayboga 2015).

These occurrences have underlined forecasts that the Ilisu Dam would militarise and lead to human rights violations. The Turkish state has been called to accept responsibility however they have insisted on going forward with the project. The economic and political implications for Iraq need to be highlighted as it is dependent on the affected Tigris River and the ongoing war within the country has also shown that large water infrastructures intensify existing conflicts. The establishment of a healthy negotiating environment is therefore of prime importance and scientific discourses needs to be implemented on all levels to make sure that another war is averted within the region.

6.5 Future Climate Change and Water Conflicts

As discussed in previous chapters, global climate change may lead to negative changes in international water security in certain regions. Current climate change projections indicate an increase in droughts in certain regions of the world and increase in floods in others. The world will also experience accelerating variability in terms of timing, amount and aerial distribution of rainfall. These associated stressors with climate change may cause or worsen local violence and political actions related to securing water for their growing populations and economies as well as ensuring suitable food supply especially crops dependant on irrigation.

The warming of the world will drive conflict and instability on local, regional and global scales. Climate and environmental change have consequently been labelled as clear threats to global security which cannot be ignored. The most likely regions to suffer from water-related security conflicts have been identified as Central and South Asia due to the unique combination of rural economies which are dependent on single annual weather events such as monsoons, numerous cross-boundary/border rivers which flow from tropical glaciers in the Himalayas as well as the continued rapidly growing human population. Environmental pressures are layered in regions who have characterised borders, extended national antagonisms as well as open conflict. Other regions over the world will also likely suffer from water stress related to climate change, however, Central and South Asia are the only regions with the combination of these mentioned factors (Bhalla 2018).

As shown in previous case studies, major rivers which flow through arid regions hold an increased risk for interstate conflict between upstream and downstream states. Upstream countries which control headwaters are motivated to store water as a scarce and valuable resource while downstream countries have incentives to enforce their agenda on weaker neighbours. The Nile, Tigris, Euphrates, Mekong and the Indus

Rivers all meet the set criteria, have been identified as possible areas of conflict and in some cases conflict has already been ongoing. The case studies provided in previous sections have clearly shown that these rivers are indeed regions of conflict which may intensify with continued political instability or differences as well as stressors accompanied by climate change. Climate change will heighten the pressure on disputes and it is therefore of great importance that governments seek mutual cooperation and international institutions support cooperation (UNEP 2008; McKie 2015).

The combination of water insecurity and urbanisation, migration, pollution as well as radicalisation and proliferation of small arms may easily result in conflict. Examples of areas characterised by high vulnerability which has descended into regional conflicts include Darfur, Yemen, Ethiopia and Burkina Faso as well as Afghanistan and Pakistan. Water may not be the sole cause of conflict as the socio-economic conditions, water shortages or imbalances of water distribution are also drivers for conflict especially in marginalised societies.

"Water wars" which are defined by conflicts which are driven by water issues alone are considered to be unlikely but water disputes are already a reality across the globe especially in arid regions. Water disputes or water riots are therefore a more appropriate term than "water wars". Small-scale riots have occurred among farmers in China, Ethiopia and Egypt as well as Central America and have been similar to food-related riots which erupted around the world in 2008. Regions which are of significant risk are those who are characterised by strong ethnic or tribal divisions. Water riots may consequently drive towards the marginalisation of parts of society away from water stress. This migration will not just be limited to poor countries. The desertification of Mexico and Central America which has placed increased pressure along the USA-Mexican border clearly shows that developed countries will also be affected (Munguía 2006; Kreamer 2012; McKie 2015).

An indirect link, therefore, exists between climate change impacts on the environment and migration. Impacts of climate change may initially cause violent conflict and consequently force affected people to flee due to violence. Conflict will then further deteriorate the environment and accelerate environmental degradation in the region which will lead to further migration. Drought has considered to being one of the climate change impacts which could trigger these conflicts. A potential for conflict therefore exists when a population experiences social discrimination in terms of access to safe and clean water and when water scarcity contributes to distributional conflict.

The impacts of water scarcity and soil degradation on food security have mainly led to migration instead of violent conflict. Migration has consequently also been labelled as one of the most distressing effects of climate change (Campbell et al. 2007). Migration in combination with politicisation of ethnicity, financial effects of displacement as well as expanding of conflicts could worsen current conflicts. Migration will also place increased pressure on the environment where they relocate due to a sudden increase in demand for resources such as water and food. The resultant competition between groups over limited resources may cause conflict if there is a history of tension between these groups due to social and cultural differences (Hermans 2012).

Climate-induced water stress is and will become an increased prime challenge in the twenty-first century. A comprehensive approach will be needed as climate change is intertwined with numerous factors as mentioned previously in this chapter. Current treaties need to be designed better and be more flexible to ensure that the different levels of water insecurity or stress as well as specific needs are considered for individual countries. It should also be a mechanism to protect weaker countries which lack leverage to deal with hegemons directly. A win–win situation, therefore, should be created for all riparian states and leadership needs to be encouraged in regions of high risks of water conflicts or disputes.

6.6 Water as a Global Risk

The previous sections clearly indicated and emphasised the role of increased water stress, continued civil unrest, competing interests as well as future climate change in the creation of water disputes as well as environmental refugees. Water can, therefore, be deemed to be a risk on all levels, i.e. local, regional and global if risks and vulnerabilities are not identified together with the implementation of adaptations and responses to build resilience.

Failure to implement environmental, social and economic measures to try and combat effects of water stress, continued droughts, as well as future impacts of climate change, will lead to further environmental degradation, drive possible mobilisations of dissent between water users and cause subsequent political unrest in extreme cases. This will in turn also lead to the creation of environmental refugees which will place further strain on neighbouring countries by deepening water pollution and scarcity. Continued poor water management practices as well as ill-maintained water infrastructure will exacerbate water scarcity problems even further and increase water-related risks in all sectors.

Current water-related problems, therefore, need to be identified to establish the costs thereof for the environment, social and economic spheres. The future impacts of climate change also need to be considered as this will also play a major role in the creation or deepening of vulnerabilities as well as risks within regions. The following sections will look at the current water-related risks as well as their costs to evaluate water as a global risk.

6.6 Water as a Global Risk

6.6.1 Costs of Lack of Clean and Reliable Water

It is estimated that approximately 900 million people around the world do not have access to improved sources of clean drinking water and that 2.6 billion of the world's population do not have access to improved sanitation. Access to water, sanitation and hygiene is a human right, yet billions are still faced with daily challenges accessing even the most basic of services.

Without any intervention, by the year 2025, as much as two-thirds of the world's population will be living under serious water shortage conditions and one-third of the population will be living in regions experiencing water scarcities. The year 2050 will see half of the world's population living in conditions of absolute water scarcity (Curry 2010; Ribolzi et al. 2011). Numerous countries are either suffering from intense water shortages or running the risk of being added to the water scarcity statistics. Water scarcity is, therefore, a major global risk as indicated in this and previous chapters and requires attention on all scales and in all water use sectors.

Rapid urban development especially within the developing world has resulted in many informal settlements which lack domestic waste disposal, sanitation as well as sewerage/effluent systems and limited water supply. People are continued to be forced to live in areas of inadequate access to sanitation and have forced them to obtain water from very limited areas. The surface water bodies within most of these areas are highly impacted and shallow wells are often near pit latrines. More than 80% of sewage waste in developing countries are estimated to be discharged untreated into water bodies which affect not only drinking water but also ecosystems which can not survive in eutrophic conditions. Continued urbanisation which is driven by a combination of external factors will therefore lead to continued water degradation and ultimately increased water scarcity within these regions as water will be unsuitable for use. It, therefore, needs to be noted that continued rapid urban development needs to be properly managed to reduce the risk of increasing water stress.

The poor population, particularly children, are mostly hurt or affected by unhygienic and insufficient water and it is estimated that every year inadequate sanitation with the lack of hygiene have claimed the lives of 2.2 million children under the age of five and is an equivalent to 6,000 deaths per day. The impact of diarrheal diseases in children under the age of 15 is greater than the combined impact of HIV/AIDS, malaria and tuberculosis. Additional global statistics related to access to clean water and water-related diseases include the following. Approximately 2.6 billion people have gained access since 1990 however 663 million are still without access to improved drinking water resources. A minimum of 1.8 billion people around the world drink from water sources which are faecally contaminated. Lastly, the proportion of the global population using improved drinking water resources increased from 76% in 1990 to 91% in 2015. Water degradation, therefore, has immense socio-economic costs as it increases health risks and costs as a whole placing strain on developing countries' and/or region's current and future economic development.

The investment in the improvement of water supply, sanitation, hygiene and water management practices can reduce the world's disease burden by approximately 10%.

Increased good hygiene behaviours can be used to combat ill effects of diseases caused by poor personal hygiene and skin or eye contact with contaminated water. Organisations attempt to address these issues by combining improved sanitation facilities with hygiene education on schools. SuperAmma is a good example of such a programme which has been successfully implemented in rural India where instead of using health messaging, emotional motivators were used to improve hand-washing behaviours. The campaign increased hand washing with soap by 31%.

For regions or countries to be able to achieve the set SDGs, proper water and sanitation will be the key foundation. Improved water and sanitation also serves as a basis for achieving gender equality as well as good health in regions. Sustainable water management practices will enable the improved management of food and energy production and ultimately contribute to further economic growth. Likewise, water ecosystems will also be preserved together with their biodiversity and action can be taken on future climate change by building resilience to future changes.

Ignoring the need for improved infrastructure and sustainable water management practices will lead to the continued death of millions of people per year, further losses in biodiversity and ecosystem resilience and ultimately undermine success and a sustainable future. Governments should therefore keep on being held accountable by civil society organisations and investments need to be made into water research and development. The inclusion of women, youth and indigenous communities in water resource governance should be encouraged. Lastly, awareness needs to be generated of these roles and turn them into actions which lead to win–win results as well as improved sustainability and integrity for human and ecological systems.

6.6.2 Costs of Insufficient Supply and Future Water Scarcity

Water scarcity affects more than 40% of the global population and approximately 1.7 billion people live in river basins where water use exceeds recharge and these statistics are projected to rise. Additionally, the world's freshwater supplies are deteriorating at a rapid rate accompanied with numerous threats to maintaining supplies for the rapidly increasing human population.

Both fresh surface water and groundwater resources are increasingly being depleted by mismanagement and by over-tapping, especially in countries where the natural water supply stores are less than the demand for water. Water degradation through numerous different pollution sources are also contributing to the depletion of freshwater supplies, with this problem being the greatest in countries where water regulations or enforcement is absent. The pollution of both surface and groundwater, limits the quality of water, especially in developing countries where approximately 95% of all untreated urban sewage is discharged directly into surface water bodies. The dumping of untreated urban sewage is not only occurring in developing countries. Some well-developed countries such as the USA are also guilty of this charge. Approximately 37% of all lakes in the USA are unfit for swimming as a result of this type of pollution (UN 2003; Cassardo and Jones 2011; Ribolzi et al. 2011).

Human wastes, fertilisers, pesticides, eroded soil sediments, as well as untreated waste water from industries, are among the greatest sources of pollution. The pollution of surface water results not only in the water being unsuitable for human consumption but is also applicable to crops. By ignoring these problems and not addressing them, water shortages could be aggravated, humans and ecosystems be threatened and the levels of rivers and lakes could drop significantly, not only in developing countries but also in countries across the world (UN 2003; Cassardo and Jones 2011).

A significant cost of a lack of clean and reliable water supply due to increased water scarcity or pollution is related to food security. Food security is closely linked with water as global agricultural water use accounts for 70% of all water consumption compared to 20% for industry and 10% of domestic use and many forms of energy production also requires reliable water resources. The competition for adequate supply of clean water within communities have as a result also exacerbated public and private discord and have in some cases led to increased water disputes in many regions across the world.

Scarce water resources also have an impact in terms of day-to-day cost on a personal human level mainly in the form of time spent collecting water. Millions of children, especially girls, spend several hours a day collecting water causing them not to be able to attend school. It is estimated that an additional 443 million school days are lost per annum from water-related illnesses and the associated economic losses are linked to increased health expenses, absenteeism and decline in productivity. The world's poorest countries are the hardest hit with Sub-Saharan Africa estimated to have lost 5% of its GDP in 2003 or approximately $28.4 billion annually to water-related diseases which account for more than the total debt relief and aid to the region within that year. The World Bank estimates that 6.4% of India's GDP is lost due to adverse economic impacts and costs of inadequate sanitation.

Serious problems which arise from inadequate water supply can continue for generations. An example of this has been the estimated 100,000–250,000 human deaths and the perish of millions of herd animals during the drought of 1968–1975 in Sub-Sahelian Africa which resulted in societal upheaval, significant shifts in population (displacement of 5.5 million), many children suffering from brain damage due to inadequate nutrition and the economy of eight countries devastated for decades after.

Furthermore, 20% of the world's potable water is lost from distribution pipes on account of inefficiencies (e.g. leaking pipes). Countries such as Bulgaria and Hungary have unacceptably large numbers as they lose 50 and 35% of potable water through pipe inefficiencies respectively. The Indian city of Bangalore loses approximately half of its pumped water even before it reaches the city's distribution systems. However, it is not only the developing nations that are feeling the effects of water shortages. Continued development and the intensification of the urbanisation process in developing countries, as well as the overall increase in the world's population, will cause a dramatic increase in water usage (Sikdar 2007; Ribolzi et al. 2011).

There is no substitute for water as it is an essential element for the survival of human beings and the environment. Industries and national economies are all dependent on this resource. The demand for already overcommitted national and international water resources in numerous countries has rapidly increased, especially in many of the world's largest cities. The result has been the ripple effect of disputes among riparian communities which eventually escalates into serious regional and international security issues (Shen et al. 2008; Frederiksen 2009). The increase in the world's human population is therefore not the only cause of water scarcity. Political power, policies and socio-economic relations can also induce water scarcities as a result of unbalanced power relations, poverty and inequalities (Kummu et al. 2010; UNDP 2006). Political conflicts over water especially in the Middle East have strained international relations amongst these already water-starved nations and the combination of political conflicts and the continuous increase in population will exacerbate and possibly spread these problems pertaining to water even further (O'Brian and Leichenko 2003; Cassardo and Jones 2011).

Furthermore, increasing scarcities owing to climate change, population growth, industrialisation, inefficient agricultural practices, and the degradation and maldistribution of water resources, could also compound already tense interstate relations and could cause mass migrations of environmental refugees (Guslits 2011). These mass migrations will place immense strain on the water resources within these affected regions or countries and might lead to social conflict and disputes, increased pressure on food security and place pressure on future economic growth as a result of increased competition between the human population and primary water use sectors.

It is therefore very clear and definite that these externally imposed stresses can lead to social unrest and disputes, political instability and in some cases as indicated previously armed conflict. Global estimates of the number of people living in areas with water stress therefore differ significantly between studies (Vörösmarty et al. 2000; Alcamo et al. 2003, 2007; Arnell 2004). Climate change is only one of many factors that influence future water stress; demographic, socio-economic and technological changes possibly play more important roles in most regions. The number of people living in water-stressed river basins would increase significantly mainly due to population growth projections.

6.7 Managing Water Tensions, Risks and Conclusions

It was estimated in 2010 that approximately 80% of the world's population were living in areas where water supply was not secure with leakages in the water supply system being a predominant vulnerability in both developing and developed countries. The contamination of drinking water or sabotage of distribution chains has also been regarded as likely terrorist acts even though the quantity of toxic chemical needed to contaminate municipal water supply is hard to define. Most biological pathogens fail to survive in water as a contaminant in the developed world due to standard water treatment practices however the vulnerability of water supply post-treatment has increased in risk. Consequently, some developed nations such as the USA have started to develop new safety measures such as remote monitoring technologies as well as critical system redundancies and have adopted new protocols to fortify water supplies and to decrease these vulnerabilities and risks.

Conflicts over water resources have been described to be more complicated than a direct parallel to other resources. Various distinctions should be made between the types of resources and the paths by which they might contribute to possible conflict. The complication nature of water conflicts is further emphasised by water not being lootable like other conflict resources such as diamonds, not being geographically fixed as well as being intimately linked to livelihoods, local cohesion as well as political tensions.

The continued disputes between India and Pakistan have become an example of how listed water disputes serve as an impediment to cooperation between countries. India's growing energy needs for continued economic growth have increasingly clashed with the farmers in Pakistan who are dependent on shared waters. India's continued construction of several dams in the Indus River basin has led to Pakistani military and jihadi groups identifying water disputes as a core issue which needs to be resolved for the relations between these two countries to normalise.

The monitoring and assessment of the globe's water quality have been described as an essential part in trying to understand the intensity and scope of the world's water quality challenge. The coverage of data is however inadequate in many parts of the world and an urgent venture focussed upon the expansion of water quality data collection, distribution and analysis will be needed to try and identify hot spot areas of water pollution which can be deemed as high-risk areas.

The present situation regarding water planning and management holds many challenges. The future thereof will become even more challenging as a result of an increase in the demands of the world's population and the increased impacts that will be associated with climate change. It is estimated that the world's population will increase by 2.5 billion people in 2050 to bring the world's total population to 9.2 billion. This increase in population is equivalent to the total number of people that were in the world in 1950. Most of the growth in population will occur in the less-developed regions. Most of these regions are already the most severely affected by the lack of sanitation and drinking water services so that an increase in population will only aggravate the situation.

Growing population pressures, unsustainable consumption as well as accompanied escalating environmental stresses have led to the escalation of strains on freshwater resources around the globe. The case studies included in this chapter are prime examples of how these pressures have put mounting strains on the environment but also on affected human populations and political relations.

Many river basins around the world are increasingly considered closed, meaning that the renewable water within the basin has already been allocated to various water sectors and the environment itself with little or no spare capacity. Examples of such river basins include from the Nile to the Tigris-Euphrates to the Indus as well as the Colorado River (USA) and the Yellow River (China). Closed river basins and rivers are not able to absorb new water demands or buffer fluctuations in supply. Changes in water use in one part of the system will therefore echo to users elsewhere in the basin, increasing pressure on consumers, the environment and policy makers.

It is estimated that more than 1.4 billion people currently live within closed river basins. The constant increase in water demands of primary water sectors will continue to increase the number of closed basins around the world and risk outrunning sustainable supplies in many more regions. Currently, projections show that the world's water requirements will exceed renewable resources by 40% in 2030 if we do not consider and implement extensive efficiency gains as well as policy improvements. China's water supplies are estimated to only fulfil three-quarters of its demand and India only half by 2030.

Global climate change will further aggravate these challenges by shifting rainfall patterns threatening to reduce water availability in some regions on the one hand and inflicting extreme weather events such as stronger storms in others and increasing potential droughts and floods on the other. The populations within regions experiencing drought will migrate to wherever resources are and as a consequence often triggers conflict. With the constant increase of water scarcity around the world, conflicts over resources have become common and water-related conflicts will definitely increase.

For the world to meet its growing water needs it will require more effective use of available water resources and ultimately enhanced collaboration between water use sectors and communities within nations as well as between countries in international basins. Water policy changes made by one user can ultimately affect the timing, location and amount of available water to other users as consumers are intimately linked by these shared water resources. As shown in the Nile confrontation case study, confrontation is not a productive way to navigate trade-offs as conflict can not assist in meeting increased water demands. Due to the indistinguishable ties between various water users in shared basins, only cooperation between nations and water users will be able to achieve the identification and implementation of necessary trade-offs and assist in meeting rising water demands. Water resources ignore political boundaries and water managers will need to learn how to manage shared water supplies as allies instead of rivals or more and more regions will suffer from increased water shortages. Increased water scarcity around the globe will cause politics to play a very powerful role in deciding water allocations across and within borders. Strong institutions will therefore be needed which should include equitable economic reallocation schemes, communication management as well as legislation to manage water tensions. The history of cooperation does not show much promise for the future and more focus needs to be put on water disputes which are not bothered by international borders or diplomacy. The establishment of a healthy negotiating environment and a change in the prevailing discourse from a political to a scientific level will be required on all scales to ensure the development of adaptations and responses which improve resilience and minimise of vulnerabilities.

References

Alcamo J, Doll P, Henrichs T, Kaspar F, Lehner B, Rosch T, Siebert S (2003) Global estimates of water withdrawals and availability under current and future "business-as-usual" conditions. Hydrol Sci 48:339–348

Alcamo J, Florke M, Marker M (2007) Future long-term changes in global water resources driven by socio-economic and climatic changes. Hydrol Sci J 52:247–275

Aleem Z (2015) Why water shortages are the greatest threats to global security. Available via http://mic.com/articles/111644/whywatershortagesarethegreatestthreattoglobalsecurity#.dBjhX1ok5. Accessed on 26 Feb 2018

Al-Masri A (2014) Water wars directed against Syria and Iraq: Turkey's control of the euphrates river. Available via https://www.globalresearch.ca/water-wars-directed-against-syria-and-iraq-turkeys-control-of-the-euphrates-river/5389357. Accessed on 26 Feb 2018

Al-Muqdadi SW, Omer MF, Abo R, Naghshineh A (2016) Dispute over water resource management—Iraq and Turkey. J Environ Prot 7:1096–1103

Arnell N (2004) Climate change and global water resources. SRES emissions and socio-economic scenarios. Global Environ Change 14:31–52

Ayboga E (2015) Ilisu Dam construction site militarized. Available via http://www.hasankeyfgirisimi.net/?p=41. Accessed on 26 Feb 2018

Bari SA (2016) The water crises in Turkey, Syria and Iraq. Available via https://www.pakistantoday.com.pk/2016/07/17/the-water-crises-in-turkey-syria-and-iraq/. Accessed on 26 Feb 2018

Bhalla N (2018) World has not woken up to water crisis caused by climate change. Available via https://www.scientificamerican.com/article/world-has-not-woken-up-to-water-crisis-caused-by-climate-change/. Accessed on 26 Feb 2018

Campbell K et al (2007) The age of consequences: the foreign policy and national security implications of global climate change. Centre for Strategic and International Studies (CSIS) and Centre for a New American Security (CNAS). Available via http://csis.org/files/media/csis/pubs/071105_ageofconsequences.pdf

Cassardo C, Jones JAA (2011) Managing water in a changing world. Water 3:618–628

Cooley H, Donnolly K, Phunisamban R, Subramanian M (2015) Impacts of California's ongoing drought: agriculture. Available via http://pacinst.org/publication/impactsofcaliforniasongoingdroughtagriculture/. Accessed on 26 Feb 2018

Cruickshank M (2013) Yemen is on the verge of running out of water. Available via https://thinkprogress.org/yemen-humanitarian-crisis-water-54a9c0b52831/. Accessed on 26 Feb 2018

Curry E (2010) Water scarcity and the recognition of the human right to safe freshwater. Northwestern J Int Hum Rights 9:103–121

FAO (Food and Agriculture Organisation) (2012) Syrian Arab Republic Joint Rapid Food Security Needs Assessment (JRFSNA). FAO Report, 26 pp. Available online at http://www.fao.org/giews/english/otherpub/JRFSNA_Syrian2012.pdf

Femia F, Werrell C (2013) Syria: climate change, drought, and social unrest. The Center for Climate and Security. Available online at http://climateandsecurity.org/2012/02/29/syria-climate-change-drought-and-social-unrest/

Frederiksen HD (2009) The world water crisis and international security. Middle East Policy 16:76–89

Frenken K (2009) Irrigation in the Middle East region in figures: AQUASTAT survey—2008. FAO Water Report, 34, 402 pp. Available online at http://www.fao.org/docrep/012/i0936e/i0936e00.htm

Gleick P (2014) Water, drought, climate change, and conflict in Syria. Am Meteorol Soc 6:331–340

Guslits B (2011) The war on water: international water security. Political Science Department. University of Western Ontario

Hermans L (2012) Climate change, water stress, conflict and migration. International Hydrological Programme of UNESCO. Available via https://www.unesco.nl/sites/default/files/dossier/climate_change_water_stress_conflict_and_migration_0.pdf. Accessed on 26 Feb 2018

Intelligence Community Assessment (2012) Global water security. ICA 2012–08. Available via https://www.dni.gov/files/documents/Special%20Report_ICA%20Global%20Water%20Security.pdf. Accessed on 26 Feb 2018

Johnson R (2015) California's drought and the politics of inequality. Available via http://www.truth-out.org/news/item/32143-california-s-drought-and-the-politics-of-inequality. Accessed on 26 Feb 2018

Jones A (2017) Food security: how drought and rising prices led to conflict in Syria. Available via https://theconversation.com/food-security-how-drought-and-rising-prices-led-to-conflict-in-syria-71539. Accessed on 26 Feb 2018

Kahn D (2015) Drought: political temperatures rise over conflicts between urban and rural water use in California. Available via http://www.eenews.net/stories/1060016956. Accessed on 26 Feb 2018

Kasler D, Reese P (2015) California drought impact pegged at $2.7 billion. Available via http://www.sacbee.com/news/state/california/wateranddrought/article31396805.html. Accessed on 26 Feb 2018

Kreamer DK (2012) The past, present, and future of water conflict and international security. J Contemp Water Res Educ 149:87–95

Kummu M, Ward PJ, de Moel H, Varls O (2010) Is physical water scarcity a new phenomenon? Global assessment of water shortage over the last two millennia. Environ Res Lett 5:1–10

Lopour J (2016) Yemen, water and conflict. Available via https://www.cigionline.org/articles/yemen-water-and-conflict. Accessed on 26 Feb 2018

Lopour J (2017) Yemen, water, conflict and cholera. Available via https://reliefweb.int/report/yemen/yemen-water-conflict-and-cholera. Accessed on 26 Feb 2018

Lumaye S, Wuthnow J, Samaranayake N (2016) China and India's slow-moving path to 'Water Wars'. Available via http://nationalinterest.org/feature/china-indias-slow-moving-path-water-wars-18254. Accessed on 26 Feb 2018

McKie R (2015) Why fresh water shortages will cause the next great global crisis. Available via https://www.theguardian.com/environment/2015/mar/08/how-water-shortages-lead-food-crises-conflicts. Accessed on 26 Feb 2018

Mhanna W (2013) Syria's climate crisis. Available via http://www.al-monitor.com/pulse/politics/2013/12/syriandrought-and-politics.html#. Accessed on 26 Feb 2018

Michel D (2013) Egypt, Ethiopia water dispute threatens nations. Available via http://www.ibtimes.com/egypt-ethiopia-water-dispute-threatens-nations-1324189. Accessed on 26 Feb 2018

Mohtadi S (2013) Climate change and the Syrian uprising. Available via http://thebulletin.org/web-edition/features/climate-change-and-the-syrian-uprising. Accessed on 26 Feb 2018

Muhammad J (2010) Scarce water the root cause of Darfur conflict? Available via http://www.finalcall.com/artman/publish/World_News_3/article_6808.shtml. Accessed on 26 Feb 2018

Munguía VS (2006) Water conflict between the US and Mexico: lining of the All-American Canal. Human Development Report 2006, UNDP

Ng D, Mukherjee T (2017) As a river dies: India could be facing its 'greatest human catastrophe' ever. Available via https://www.channelnewsasia.com/news/cnainsider/as-a-river-dies-india-could-be-facing-its-greatest-human-9060070. Accessed on 26 Feb 2018

O'Brian Kl, Leichenko RL (2003) Winners and losers in the context of global change. Ann Assoc Am Geogr 93:89–103

Patranobis S (2016) China blocks Brahmaputra tributary, impact on water flow in India not clear. Available via https://www.hindustantimes.com/india-news/china-blocks-brahmaputra-tributary-impact-on-water-flow-in-india-not-clear/story-QVAYbO2iOBFUSynwwpyneN.html Accessed on 26 Feb 2018

Pedroncelli R (2015) Tiny endangered fish highlights California drought conflicts. Available via http://www.cbsnews.com/news/tinyendangeredfishhighlightscaliforniadroughtconflicts/. Accessed on 26 Feb 2018

References

Pimentel D, Whitecraft M, Scott ZR, Zhao L, Satkiewicz P, Scott TJ, Phillips J, Szimak D, Singh G, Gonzalez DO, Moe TL (2010) Will limited land, water, and energy control human population numbers in the future? Hum Ecol 38:599–611

Polgreen L (2007) A godsend for Darfur, or a curse? Available via http://www.nytimes.com/2007/07/22/news/22iht-22polgreen.6764928.html. Accessed on 26 Feb 2018

Poppick L (2014) California droughts could have dangerous ripple effects. Available via http://www.livescience.com/49287californiadroughtsrippleeffects.html. Accessed on 26 Feb 2018

Ramachandran S (2015) Water wars: China, India and the Great Dam Rush. Available via https://thediplomat.com/2015/04/water-wars-china-india-and-the-great-dam-rush/. Accessed on 26 Feb 2018

Ribolzi O, Cuny J, Sengsoulichanh P, Mousque's C, Soulileuth B, Pierret A, Huon S, Sengtaheuanghoung O (2011) Land use and water quality along a Mekong tributary in Northern Lao P.D.R. Environ Manag 47:291–302

Saleeby S (2012) Sowing the seeds of dissent: economic grievances and the Syrian social contract's unraveling. Available via http://www.jadaliyya.com/pages/index/4383/sowing-the-seeds-of-dissent_economic-grievances-an. Accessed on 26 Feb 2018

Schlein L (2011) Water scarcity root of Darfur conflict. Available via https://www.voanews.com/a/water-scarcity-root-of-darfur-conflict-123688459/158292.html. Accessed on 26 Feb 2018

Shen Y, Oki T, Utsumi N, Kanae S, Hanasaki N (2008) Projection of future world water resources under SRES scenarios: water withdrawal. Hydrol Sci 53:11–33

Sikdar SK (2007) Water, water everywhere, not a drop to drink. Clean Technol Environ Policy 9:1–2

Swanson D (2014) Water and conflict. Available via http://www.irinnews.org/analysis/2014/04/22/water-and-conflict. Accessed on 26 Feb 2018

UN (United Nations) (2003) The United Nations Development Report: water for people, water for life. Executive Summary, UN

UNDP (United Nations Development Programme) (2006) Human development report 2006: beyond scarcity: power, poverty and the global water crisis. United Nations Development Programme, New York

UNEP (United Nations Environmental Programme) (2008) Vital water graphics: an overview of the state of the world's fresh and marine waters, 2nd edn. UNEP/GRID-ARENDAL

Vira B (2016) Droughts and floods: India's water crises demand more than grand project. Available via https://theconversation.com/droughts-and-floods-indias-water-crises-demand-more-than-grand-projects-60206. Accessed on 26 Feb 2018

Vörösmarty CJ, Green P, Salisbury J, Lammers RB (2000) Global water resources: vulnerability from climate change and population growth. Science 289:284–288

Walker T (2015) California drought: almond growers fight back over reports they are causing chronic water shortages. Available via http://www.independent.co.uk/news/world/americas/california-drought-almond-growers-fight-back-over-reports-they-are-causing-chronic-water-shortages-10224339.html. Accessed on 26 Feb 2018

Whitehead F (2015) Water scarcity in Yemen: the country's forgotten conflict. Available via https://www.theguardian.com/global-development-professionals-network/2015/apr/02/water-scarcity-yemen-conflict. Accessed on 26 Feb 2018

Whittington D (2016) Nile valley water conflict: can Egypt live with Ethiopia's grand renaissance dam? Available via https://www.juancole.com/2016/06/conflict-ethiopias-renaissance.html. Accessed on 26 Feb 2018

Wines M (2014) West's drought and growth intensify conflict over water rights. Available via https://www.nytimes.com/2014/03/17/us/wests-drought-and-growth-intensify-conflict-over-water-rights.html. Accessed on 26 Feb 2018

Worldwatch Institute (2018) Desertification as a source of conflict in Darfur. Available via http://www.worldwatch.org/node/5173. Accessed on 26 Feb 2018

Wuthnow J (2016) Water war: this river could sink China-India relations. Available via http://nationalinterest.org/feature/water-war-river-could-sink-china-india-relations-15829. Accessed on 26 Feb 2018

Part II
Water as a Regional and National Risk

Chapter 7
Evaluation of Southern and South Africa's Freshwater Resources

The southern African region experiences a great variance in climate as well as in terms of water resources. Water resources are characterised by great temporal variability especially in drier countries which are in the southernmost part of the region. The region has a high number of transboundary rivers and is importantly indicating a cooperation pattern rather than conflict over water as an increasing scarce resource is at variance of the global norm. This can be attributed to more robust institutional development of transboundary river basins than expected in the region.

Both surface and groundwater of southern Africa and specifically South Africa are increasingly under pressure due to the constant growth in the population as well as increased climate variability. It is important to note that more than half of the country's WMAs are in a water deficit after the allocation of enough water for rivers and environmental flow requirements despite significant water transfers into the country from other systems to assist in meeting water requirements. Both surface and groundwater resources are continually being degraded by all water sectors in various manners and have given rise to multiple water quality issues which affect water availability as a result. High water losses as well as overall increased water scarcity and stress within the country have led to the implementation of widespread water restrictions which have been put in place throughout the country with the recent (2015–2017) drought. For the country to improve on its freshwater resources, it needs to improve or adapt its overall management approaches and invest in treatment technologies.

7.1 Background to Southern Africa's Freshwater Resources

As emphasised in previous chapters, water is a strategic resource which is in flux and a fugitive of nature unlike other resources such as oil, gold, coal, etc. which are stock resources. The flux and fugitive nature of water means that the special resource can become a renewable resource if it is managed correctly. The distribution of water is determined by physical elements, i.e. the hydrological cycle, which in turn determines the spatial and temporal distribution of water over southern Africa and the rest of the world. The following sections focus on the water availability within the region, the identified vulnerabilities, challenges and impacts of its water resources as well as look at possible conflicts or cooperation with future water stress or scarcity.

7.1.1 Water Availability

The Southern African Development Community (SADC) ranges from 35° of the northernmost regions of the Democratic Republic of Congo (DRC) to the southernmost point of the Cape Agulhas characterised by vast differences in hydrology and climate. The southern African region is characterised by large differences between countries in terms of climate, hydrological conditions and governmental arrangements even though similarities exist. The northern regions located slightly north of the equator are humid and well-watered and in direct contrast to the deserts located along the Namibian coastline. The climate of the region is also profoundly influenced by the two ocean currents passing on either side. The Agulhas current on the east coast brings warm water from the northern region, producing warm and humid climate along the eastern coastline while the cold Benguela brings cold water from the Antarctic into the south-western coast of the region and produces cooler and drier climate along the western coastline. The national average depth of rainfall in the region ranges from as low as 285 mm per annum in Namibia to 1,543 mm per annum in the DRC (FAO 2006).

Southern Africa is therefore characterised by uneven spatial distribution of rainfall with a steep gradient from north to south and east to west. The uneven distribution of rainfall across the region potentially leaves the most economically diverse countries (South Africa, Botswana, Namibia and Zimbabwe) all on the "wrong" side of the global rainfall average of 860 mm per annum.

The region experiences a great variance in climate as well as in terms of water resources which is also characterised by great temporal variability especially in drier countries located in the southernmost part of the region. Years of drought broken up my large-scale floods with many few years of receiving the supposed average rainfall with year-on-year variances around the long-term norm of as high as 30–35% is typical for this region (Hirji et al. 2002).

Large-scale water storage and transfer infrastructure have consequently been developed due to constant spatial and temporal variability and uneven distribution of rainfall. South Africa and Zimbabwe are ranked as amongst the top dam builders in the world as very few water bodies are underdeveloped. Most of these developed water bodies within these two countries are used beyond sustainable supply rate especially when viewed with environmental water requirements (WCD 2000).

"New" water for growing economic and social uses in these countries will require the more efficient use of current water supplies through Water Demand Management (WDM) measures which will require significant investments in water-efficient technologies especially in terms of agricultural and industrial production processes as well as in other water use sectors such as domestic use with the adoption of water-wise devices. The development of more international water transfer schemes in the region is seen as an alternative but these schemes are usually accompanied with a range of environmental, social, economic and political implications which could make it a more complicated alternative. Other important hydrological factors which characterise the region and which should also be considered include the high number of transboundary rivers as well as the region's high reliance on groundwater. Five of the SADC states have a water resources dependency ratio over 50% meaning that these countries rely on water generated outside their borders to supply more than half of their total water stock (FAO 2006).

The high number of transboundary rivers will link the future of these basin states and their populations through the impacts on water quantity, quality or flow patterns being transferred downstream. The region also has a high reliance on groundwater where several of them are shared. Most of these major aquifers are not coincident with major river basins in the region and form a stock of fossil groundwater which may complicate water management in the region further due to varied flow characteristics, recharge rates, permeability and transmissivity which is not always well understood. These two factors complicate water management in the region and link the developmental futures of neighbouring states.

Therefore, developments on an upstream section of a river such as a dam for domestic water supply, irrigation or hydropower will impact downstream flow and consequently the area through environmental and socio-economic impacts.

Some of these downstream impacts include a decrease in the quantity of water available downstream and various other water-related issues which may include the following. The return flow of water used for intensive agricultural production will typically have high levels of nutrients due to intensive fertiliser and pesticide use and high salinity content which will all impact the ability of downstream ecosystems to function properly due to nutrient build-up leading to eutrophication and a general decrease in overall water quality. A decrease in water quantity downstream due to consumptive use upstream will exacerbate the situation further and may lead to an ecosystem reaching a tipping point and no longer be able to absorb and process nutrients and other pollutants being passed to it (Turton 2008).

In terms of groundwater, about 25–30% of the region is covered by major groundwater basins which are aerially extensive and show primary porosity and permeability. These basins include the Permian-Triassic Karoo sedimentary basin which covers large areas of South Africa, Botswana, Zimbabwe, Zambia, Namibia and Angola. The large tertiary sedimentary basin on the DRC and western Angola has not been exploited as it lies in a humid region. Along coastal areas, especially Mozambique and Tanzania, extensive Cenozoic coastal plain deposits which are an exploitable aquifer resource. Similar deposits occur on the western coast of the region but to a lesser extent. In terms of the Kalahari basin, it is a vast area of unconsolidated aeolian sand, tertiary to recent age, which forms a potential huge aquifer source. Some areas however in the Kalahari basin are characterised by saline groundwater.

Groundwater across the region is mostly only a perennial source of water and the transboundary nature of the aquifers are not always fully understood. The eastern and northern part of the SADC region is characterised by humid tropical and equatorial climate conditions where perennial surface water is predominantly used and groundwater receives less attention. When utilised, groundwater resources are also generally not as intensely monitored and managed. More arid parts of the region (Namibia, Botswana and northwestern South Africa) surface water is only available for short periods of time during the rainy season and groundwater has assumed great importance as it is perennially available and mostly a principal source of freshwater (Christelis et al. 2010).

Groundwater is a primary source of drinking water for human populations and livestock in the driest parts of the region and it is a main or complementary source of intensive irrigation in some parts of the region. Groundwater also plays a significant role in achieving food security through small-scale irrigation especially in dry areas. It is estimated that 70% or 250 million of the people living in southern Africa rely on groundwater as a primary source of water. Approximately 36% of the region's urban population rely on groundwater (Molapo et al. 2000) while 23% is supplied by surface water. The other 40% of the population, mainly in rural areas, remain unserved with formal water supply schemes and consequently rely on informal and traditional developed groundwater sources which include hand dug wells, springs, etc. (Braune and Xu 2008).

The importance of using groundwater for larger settlements and bigger scale water supply has often been understated. Groundwater already plays a critical role during both short-term and extended drought. Droughts are endemic to the region and with the frequency and intensity expected to increase in the future, the role of groundwater will become increasingly more critical.

There is however still a general prejudice among public and political authorities against the use of groundwater as it is now viewed as a reliable, cost-saving and clean source of water supply. Its importance for socio-economic development is therefore still poorly understood in the region. The use of groundwater in the region is therefore often under-utilised, unsustainably exploited and inadequately managed. Despite some progress in the region, the performance of integrated water resource management is still poor when compared to relevant international best practice especially in terms of institutional capacity and the establishment of an enabling environment. Awareness is present at a decision-making level regarding the importance of groundwater management but it is not yet sufficiently reflected in policies, legislation and their implementation or enforcement. The following shortcomings or trends related to the groundwater resources have been identified:

- Even though the bulk water supply is unable to satisfy remote demand at reasonable costs, there is still a general bias towards surface water resource development and long distance piped water supply systems.
- Critical shortcomings related to groundwater management primarily include the lack of an organisational framework, institutional and technical capacity as well as in terms of information and awareness at all levels.

- Lack of macro-planning which enables activities to be undertaken on an ad hoc basis or crisis response basis (Braune and Xu 2008).

The SADC Groundwater Management Programme (SADC 2005) is, however, trying to address these mentioned shortcomings at a regional level through the development of a Code of Good Practice for groundwater Development, a regional Hydrogeological map and the establishment of a Groundwater Management Institute.

The unmanaged use of this resource by competing sectors will also be a threat to those who rely on it and the expanding industry drawing down aquifer levels as well as pollution of these aquifers by agriculture and mining will add even more pressure and become a growing concern as more and more people rely on groundwater due to increased climate variability (Braune and Xu 2008).

The water resources in terms of both surface and groundwater of southern Africa are increasingly under pressure due to the constant growth in the population as well as increased climate variability. It is therefore evident that water availability is a limiting factor for future socio-economic growth and development in the region. It is consequently of great importance that the management of the resource-base is conducted in a responsible and sustainable manner and be informed by robust scientific processes.

7.1.2 Challenges and Vulnerabilities

The highly variable rainfall in combination with common high daily, monthly and annual coefficients of variation with intra-annual anomalies cause the region's river flow to be highly variable (Nicholson 2000). Furthermore, the African continent (and Australia) has the world's lowest rainfall to runoff ratio and a very low reliability of yield and changes in these factors will likely have exaggerated effects on human well-being. This once again emphasises that water is a limiting factor in the region and has led some to suggest that some of these countries are "hostage to their hydrology" (Grey and Sadoff 2007) which highlights the link between economic performance and development with rainfall and runoff regimes. Water is, therefore, an important strategic resource to the economies of the SADC region as it forms an input for various sectors and has the potential to be developed to contribute to the achievement of food security and poverty eradication.

The region's water resources have been and will continue to be developed and managed to promote agriculture, industries, mining as well as power generation to continue regional development. Agriculture is the largest water user in the region, using between 70 and 80% of available water resources. Botswana and South Africa have the lowest percentage of agricultural water use indicating that other countries will follow suit as they become more diversified and agricultural water use will be placed in competition with other sectors of the economy. Continued development of agricultural water use from small-scale or subsistence farming reliant on rainfall to irrigated agriculture will likely be impacted by rapid industrial and urban development as the water demand increases by other sectors such as mining and industrial processing in the region. Electricity production will have to increase to try and keep up with rapid development in the region which will consequently also increase water consumption. Industrial water use will mainly increase in terms of electricity production (Hirji et al. 2002). Hydropower developments are also likely to increase especially in the DRC, Zambia, Mozambique and Angola which may have various long-term environmental and social consequences (SADC 2005). Mining plays a major role in the region in terms of foreign exchange earnings and employs a large amount of people. Further development in this sector is promised to improve socio-economic development, however, the mining of minerals (gold, platinum group metals, chromites, manganese, uranium and coal) will require well-developed environmental management strategies which consider the mitigation of its associated impacts. Even though this sector is not a big water consumer, it is accompanied with a wide range of water quality issues such as acid mine drainage which have an impact on distant water users as the water passes underground.

The activities associated with the main water use sectors which include agricultural, industrial and domestic, all have the potential to pollute both surface and groundwater reserves and can in some cases contaminate these water sources to such an extent that they become unusable for other water users. In a case of it happening in one country, it proves very difficult who needs to bear the costs for the contamination and this becomes even more complex on a transboundary level.

The uneven spatial and temporal distribution of rainfall over the region may also potentially limit future economic growth because of increasing water supply insecurity. The four countries which will be affected most by this will be the four most economically developed countries namely, South Africa, Botswana, Namibia and Zimbabwe as they all fall on the "wrong" side of the global average rainfall as described previously. The whole region also faces severe groundwater shortages which will jeopardise lives and livelihoods which are directly dependent on it, especially poor communities, as well as inhibit future economic growth. Groundwater also forms the basis of rural water supplies as it sustains livelihoods especially in poor communities and the pollution or unsustainable use thereof may have dire consequences for these dependent populations (Turton 2008).

Integrated Water Resource Management has been highlighted as best practice for sustainable water management as it focuses on river basins as the units of management however within the southern African context it also needs to include the following factors:

- Groundwater aquifer systems seldom corresponding with surface water management unit or river basin; and
- Groundwater systems are in most cases transboundary by nature which complicates water resource management (Turton et al. 2006).

The region has put a complex set of transboundary agreements in place which govern river basins however it lacks international groundwater treaties of a similar status which could be a cause for potential conflict in future.

The inherently low conversion rate of rainfall to runoff which affects both surface river flows and groundwater recharge, which makes major investment in infrastructure to store limited stream flow a necessity to assure supply. Insecure supply will destabilise the foundation of the modern industrial economy which most countries strive for, places pressure on all water use sectors through increasing water-related risks and ultimately affects socio-economic growth. A reduction in aquifer recharge has also been highlighted as a probability due to the nonlinear recharge of groundwater at low levels of rainfall in combination with current predictions of higher temperatures and lower rainfall. This will lead to the groundwater status of the region to become more stressed due to considerable decrease in recharge and consequently increase vulnerability of the poor which largely depend on this water source (Turton et al. 2006).

The mass migration from rural to urban regions have been associated with high rates of urbanisation together with normal urban-specific population growth. The associated economic growth within these expanding urban regions has often taken precedence over environmental protection or conservation (Biggs et al. 2008). The impacts of climate change on water resources as well as other economically natural and agricultural resources have become unescapable (Schulze 2005; Descheemaeker et al. 2010). These observed changes together with weak governance structures, political instability as well as low institutional capacities which characterise many countries in the region highlighting the great risk of widespread environmental degradation as well as the destruction of resources which sustain quality life.

The region is therefore characterised by high vulnerability to global change factors such as climate change, land use change caused by foreign investment and population growth (Parnell and Walawege 2011). These factors coupled with changes in economic practices which are driven by unexpected drivers such as high proportion of the population with HIV/AIDS, accelerate localised environmental degradation and places increase stress on water supply. To counter these effects, innovative technologies, policies and management strategies need to be designed and implemented to mitigate or minimise effects of human-induced changes on both global and local level.

The region is currently experiencing a rapid transition from subsistence livelihoods due to increased urbanisation driven by local economic drivers and in some cases being displaced from land as intensive agriculture, mining developments and other forms of land use grows. The unprecedented rate and extent of these tightly coupled transitions and changes are likely to become exacerbated by the effects of climate change and further highlights the importance of preserving natural resources, securing the sustainable use by promoting improved or optimum efficiency in the agricultural sector. The constant expansion of urban centres together with growth in intensive agriculture in the region will be accompanied with the deterioration of water quality in numerous catchments and make the provision of sanitation an increasingly bigger challenge (Bere 2007; Edwards and Withers 2008). These factors place additional pressure and strain on the already stressed water supply and reticulation systems (Ngcobo et al. 2013) as well as downstream water quality which will consequently potentially affect economic production and ultimately economic growth, food security and political stability.

7.1.3 Possible Conflicts or Cooperation with Future Water Stress/Scarcity

A coordinated approach for the utilisation and preservation of water have been recognised as being of importance and SADC states consequently signed the Protocol on Shared Watercourse Systems in 1995. The Protocol is a legally binding document which ensures the equitable sharing of water and aims to ensure efficient conservation of the scarce resource in the region. The Protocol also gives information related to the establishment, objectives, functions as well as financial and regulatory framework of River Basin Management Institutions. The countries which ratified the original SADC Shared Watercourse Systems Protocol in 2001 included Botswana, Lesotho, Malawi, Mauritius, Namibia, South Africa, Swaziland, Tanzania, Zambia and Zimbabwe. It is predicted that three or four of the SADC states will face serious water shortages in the next 20 years due to the increased water scarcity nature of the region which makes the Protocol very important in terms of cooperative sustainable water utilisation and preservation in the region (Beekman et al. 2003).

As indicated in the previous section, the southern African region is characterised by rapid urban growth, rapid transition from subsistence farming to intensive agriculture, high vulnerability to climate variability and land use changes. These factors together with high variability of rainfall over the region, high number of transboundary rivers as well as complex nature of transboundary groundwater aquifer systems place great pressure on the region's water resources which will increase due to continued economic growth, increased water demand and use as well as the effects of climate change. Increased water demand is a reality and is driven by high population growth, urbanisation, improved welfare and living conditions as well as industrial and agricultural development (Arntzen 2001).

The southern African region possesses these elements which drive increased water demand and use in the region and consequently leads to less water available per person. These increased pressures on water resources especially those of a transboundary nature may cause possible conflict between countries as they try to secure their share for their population and various economic sectors. The hot spots for actual or potential water-related conflicts in southern Africa have however been identified as possible risk areas. Largest potential for conflicts has been identified to be possibly connected with the Lesotho Highlands Water Project, the Limpopo basin as well as the Eastern Caprivi region.

7.1 Background to Southern Africa's Freshwater Resources

The linkage between conflict and natural resources have been identified in literature related to development by numerous authors. Global trends related to water and conflict have been observed as the following.

- The incidence of militarised interstate disputes in a pair of countries are usually linked to the sharing of river basins;
- Water scarcity is usually associated with conflict with the physical geography of the basin determining the key role; and
- A river forming a border is most frequently associated with conflict (Gleditsch et al. 2004).

These trends or elements are important as it coincides with many of the circumstances found in contemporary Africa (Turton 2005). The linkage appears at first to be a naturally correct assumption however the management of transboundary rivers in southern Africa have seemed to suggest the opposite position in this regard (Turton et al. 2004; Turton 2004, 2005). The joint management of water as a scarce resource have rather acted as a driver for cooperation than conflict.

The southern African pattern indicating cooperation rather than conflict over water as an increasingly scarce resource is at variance of the global norm which can be attributed to more robust institutional development of transboundary river basins than expected in the region. River basins which have more than two riparian states in terms of the global norm tend to have bilateral regimes by a ratio of 2:1 which consequently excludes all riparian states from the agreement (Conca 2006). South Africa as one of the member states of SADC is, on the contrary, a signatory of no less than 59 international freshwater agreements and basin-wide agreements exist in all basins at risk (Turton 2005). There has also been little significant tension arising from the management of various transboundary river basins which form the hydrological foundation of the SADC economy. This observation is extremely notable given the history of intense conflict associated with the Cold War era in southern Africa (Turton 2003, 2004, 2005; Turton and Earle 2005).

Present population trends, as well as patterns of water use across Africa, suggest that more countries will exceed the limits of their economically usable, land-based water resources before 2025. Regional efforts need to emphasise that the use of water resources need to ensure the sustainable long-term benefits for the people of Africa. Each country's water resource management strategy needs to be continually aligned with that of its neighbours to ensure continued cooperation and avoid conflict within the region.

7.2 Background to South Africa's Water Availability

South Africa receives an average rainfall of 450 mm per annum and is classified as a water-stressed country. The amount of rainfall varies per annum and is largely unevenly distributed and the low rainfall consequently leads to reduced levels of runoff as well as availability of surface water (Binns et al. 2001; DWA 2011a;

UNESCO 2011). The annual runoff of rivers in the country averages 49,000 million m^3 per year where 50% is yielded by mountain catchment areas which only account for 8% of the country's surface area. The per capita water availability is approximately 1,100 m^3 per year which can be utilised due to high evaporation rates and variable flows (Binns et al. 2001; StatsSA 2010). South Africa's water resources are made up of 77% surface water and 9% groundwater with the other 14% being returned flows. Water resources are therefore extremely varied and highly stressed in certain areas.

The major river basins in the country include the Nkomati, Limpopo, Maputo, Orange-Senqu, Thukela and Umbeluzi which are also shared by neighbouring countries which include Lesotho, Swaziland, Mozambique, Zimbabwe, Botswana and Namibia (Ashton et al. 2008).

South Africa has four major transboundary basins which contain 40% of its available water resources which include the following:

- Limpopo basin covering South Africa, Botswana, Zimbabwe and Mozambique;
- Komati basin covering South Africa, Swaziland and Mozambique;
- Maputo/Usuthu basin covering South Africa, Swaziland and Mozambique; and
- Orange basin located across Botswana, South Africa, Lesotho and Namibia.

The country has been divided into Water Management Areas (WMA) to facilitate water resource management and large-scale inter-basin transfers between catchments have also been established to supplement water to metropolitan areas such as Cape Town, Durban, Port Elizabeth and the Gauteng region which in some cases are located far from major water courses. The following sections will focus on the country's available surface and groundwater resources to establish where and how freshwater is available in the country and where and what demands are placed on the scarce resource for use.

7.2.1 Surface Water

As in all cases over the world, the amount of water available for human use or support of aquatic ecosystems is dependent on the availability and sustainability of the resource in the specific region or area and requires integrated management due to the interlinking of rainfall, surface flows and groundwater recharge.

One of the most important factors which should be considered in terms of water availability in the country is climate variability especially uneven rainfall distribution across the country's catchments. The northern and western parts of the country are predominantly semi-arid and receive low levels of rainfall. The following characteristics further complicate the management of the country's already scarce water resources:

- The highly seasonal rainfall and runoff;
- Long dry season and short wet seasons in many parts of the country; and

7.2 Background to South Africa's Water Availability

- High variability in annual flows which are linked to floods and droughts.

The country consequently invested in storage dams to bridge these periods of low flow and to try and assure water supplies. The country has invested in approximately 320 major dams which together have a storage capacity of 32,412 million m^3 which is equivalent to more than two-thirds of its mean annual runoff. The total dam storage capacity is therefore of a very high percentage and means that more large dams will become less efficient. Additionally, there are also thousands of smaller private farm and municipal dams which contribute further to water storage. It is important to note that dams have a major impact on aquatic ecosystem integrity and the cumulative impacts of numerous dams may be accompanied with severe effects on the state of inland waters (DWAF 2004).

It is important to note that half of the country's WMAs are in a water deficit after the allocation of enough water for rivers and environmental flow requirements despite significant water transfers into the country from other systems to assist in meeting water requirements (DWA 2010a). Therefore, any major change in rainfall or water availability will severely impact available water resources and consequently makes decreased water quality and climate change some of the primary threats to meeting future water demands. The country's water resources are already intensively used and controlled.

The quantity of water which reaches the country's rivers or natural mean annual surface runoff (MAR) is approximately 49,000 million m^3 per annum which includes water which drains naturally into the country from Lesotho (approximately 4,800 million m^3 per annum) and Swaziland (approximately 700 million m^3 per annum) (DWAF 2004). The available yield for surface water is estimated to be 10,000 million m^3 per annum. Additional water is available however the assurances are less than 2% that the water will be available in any given year. Seven of the nine provinces in the country depend on inter-basin transfers which provide half of their water requirements and a significant volume of available surface yield is also moved via these inter-basin transfer schemes to areas where requirements exceed supply. The Lesotho Highlands Water Scheme is a prime example of this. It supplies the Gauteng region through transfer from the Katse and Mohale dam in Lesotho to the Upper Vaal for the Gauteng province to have a suitable water supply to meet its constant increasing demand (DWA 2011a).

South Africa's available resources are unevenly distributed. These water sources vary across catchments and are closely linked to several conditions. The hydrological cycle, therefore, needs to be well understood to understand how much water will available and where.

The country's water sources can, therefore, be grouped into 21 areas which have been described as being the country's most important natural notional assets due to increasing water stress. These water source areas cover approximately 8% of the land area and contributes 50% to the water in our rivers. The land cover in these water source areas are predominantly natural vegetation (63%), followed by farming and forestry (28%), degraded land (3%) and mining for fuel (1%). Natural vegetation is the dominant land cover in these areas mainly due to slope and altitude having prevented intense development. Less than 1% of these water sources overlap coal deposits however the Enkangala Drakensberg and Mfolozi headwaters overlap by 30%. More concerningly is the widespread threat of acid mine drainage in the Mpumalanga Province with 50% of the province being either under prospecting or mining licence for coal which has already caused significant degradation in the province. The main rivers in each of these identified water source areas, as well as the main identified threat towards these, are shown in Table 7.1 (WWF 2013).

It is concerning that only 16% of the country's water sources are formally protected as nature reserves or parks. This should raise a red flag as these areas need to be secured and well-managed to try and ensure long-term water security. The limits for further development of surface water sources have almost been reached and prospects for spatial economic placement of new dams are rare (DWA 2010a). The costs related to transfers per cubic metre to locations where water is needed is also becoming less financially viable with longer distances and rapidly rising pump costs.

Due to 60% of South Africa's river basins including flow to or from another country, the country has entered into multiple bilateral and trilateral transboundary agreements with neighbouring states on water courses such as the Limpopo and Orange River systems. This together with inter-basin infrastructure which has enabled large-scale transfer of water between provinces have allowed the country to succeed in providing water for agriculture and industries on a large scale. The reliance on water being transferred long distances has been identified as some possible risk as political priorities in the country could lead to the reconsideration of water transfer policies. An example of this is the Water for Development and Growth Framework (2009) which makes a recommendation that inland water resources should be retained for inland rather than being transferred to coastal locations.

7.2.2 Groundwater

The groundwater of South Africa occurs in hard rocks with no pore space over approximately 90% of the country's surface where it is contained in faults, fractures and joints as well as in dolomite and limestone (dissolved openings, i.e. fissures). Groundwater occurs in primary aquifers or secondary aquifers also known as hard rock aquifers as groundwater occurs in the openings which were formed after the rock was formed. Primary aquifers are comprised of porous sediments and soils which contain groundwater in the spaces between the sand grains and are found in river sediments, coastal sand deposits as well as Kalahari deposits.

Approximately 96% of the country's borders are underlain by low-yielding aquifers and seven main transboundary aquifers. These aquifers are coupled with low demand for water due to low population density which consequently has a low risk for over-pumping or pollution disputes.

The Karoo aquifer sequence on the Kalahari is one of the major transboundary resource shared with Botswana and Namibia. Only the border with Zimbabwe is underlain to a significant extent by class 5 aquifers and those boreholes located close

7.2 Background to South Africa's Water Availability

Table 7.1 Water source areas of South Africa with main rivers and accompanied threats (WWF 2013)[a]

Water source area	Main rivers	Threats
Grootwinterhoek*	Olifants River; Klein Berg; Doring	Land degradation; climate change; alien invasive vegetation; fires
Table Mountain*	Hout; Diep	Climate change; alien invasive vegetation; fires
Boland Mountain*	Berg; Breede; Riviersonderend	Large-scale plantations; land degradation; climate change; alien invasive vegetation; fires
Langeberg*	Doring; Duiwenhoks; Naroo; Gouritz, Breede	Climate change; alien invasive vegetation; fires
Swartberg	Gamka; Sand; Dorps; Gouritz; Olifants	Climate change; alien invasive vegetation; fires
Outeniqua*	Groot Brak; Olifants	Large-scale plantations; alien invasive vegetation; fires
Kougaberg*	Kouga; Baviaanskloof; Olifants; Gamtoos; Gouritz	Climate change; alien invasive vegetation; fires
Tsitsikamma	Groot Storms; Klip; Tsitsikamma	Large-scale plantations; land degradation; alien invasive vegetation
Amatole*	Great Kei; Keiskamma; Great Fish; Tyume; Amatele	Land degradation; alien invasive vegetation; fires
Eastern Cape Drakensberg*	Mzimvubu; Orange; Bokspruit; Thina; Klein Mooi; Mthatha	Land degradation; climate change; fires
Pondoland Coast	Mzimvubu; Mngazi; Mutafufu; Msikaba	Large-scale cultivation and plantations; coal mining; land degradation
Maloti Drakensberg*	Caledon; Orange; Senqu	Large-scale cultivation; land degradation
Northern Drakensberg*	Senqu; Caledon; Thukela; Orange; Vaal	Coal mining; land degradation
Southern Drakensberg*	uMngeni, Mooi; Thugela; Mkomasi; uMzimkulu	Large-scale plantations; land degradation
Mfolozi Headwaters*	Lenjane; Black Mfolozi; Pongola	Large-scale plantations and cultivation; coal mining; land degradation
Zululand Coast	Mvoti; Thukela; Mhlatuze	Large-scale cultivation; coal mining; land degradation
Enkangala Drakensberg*	Pongola; Bivane; Assegaai; Vaal; Thukela; Wilge	Coal mining; large-scale plantations; land degradation

(continued)

Table 7.1 (continued)

Water source area	Main rivers	Threats
Mbabane Hills*	Usutu; Lusushwana; Mpuluzi; Inkomati; Pongola	Large-scale plantations; land degradation
Mpumalanga Drakensberg*	Elands; Sabie; Crocodile; Olifants	Large-scale plantations; coal mining; land degradation
Wolkberg*	Middle Letaba; Ngwabitsi; Oliphants	Large-scale plantations; land degradation; climate change
Soutpansberg*	Luvuvhu; Little Letaba; Mutale; Mutamba; Nzhelele	Large-scale plantations and cultivation; land degradation

[a]Water source areas marked with * are identified as a strategic water source area

to the border might contribute to groundwater depletion in a neighbouring country. South Africa has however international obligations under the National Water Act, to ensure that we sustainably manage a portion of the flow across our borders. The potential for disputes over transboundary groundwater is not a major concern at this point and the region can use these transboundary aquifers to improve technical cooperation, data sharing, training and research and avert potential future disputes (Cobbing et al. 2008).

The total available renewable groundwater resource in South Africa is estimated to be 10,343 million m^3 per annum or around 7,500 million m^3 per annum under drought conditions (DWA 2010b). This amount varies greatly across all water management areas where some areas have much higher groundwater reserves than others. Groundwater availability as in the case of surface water, therefore also varies across the whole country with some areas having more available than others.

Groundwater in South Africa has a relatively small contribution to the bulk water supply of the country (13%) but still represents an important and strategic water resource. In the Eastern Cape, Limpopo, Northern Cape, North West and KwaZulu Natal provinces it provides between 52 and 82% of community water supply schemes. Due to South Africa being characterised by perennial streams especially in the semi-arid to desert parts of the country, two-thirds of the country consequently largely depend on groundwater. Irrigation is still the biggest user of this resource however groundwater also supplies over 300 towns and smaller settlements (DWS 2014a).

The use of groundwater has increased dramatically over the last six decades. It is estimated that groundwater use has increased from 700 million m^3 per annum in 1950 to 1,770 million m^3 per annum in 2004 (DWA 2010b; StatsSA 2010). This water source is used for various purposes across the country. As stated previously, irrigation is the biggest user of this resource however it is also used for mining in the Highveld region and for domestic purposes especially in rural areas located in Northern, Western and Eastern Cape, KwaZulu Natal, Mpumalanga and Limpopo Provinces. Small water users are largely dependent on groundwater access and use which makes it of significant importance for livelihoods and marginalised rural communities especially in times of drought. Groundwater use in combination with adequate maintenance and

7.2 Background to South Africa's Water Availability

operation needs to be considered for future planning and management of the country's limited freshwater resources.

Contamination of groundwater resources occurs in various ways. Groundwater linked to coal deposits contain dissolved minerals which are poisonous to plants and animals. The dumping of pollution in the ground, in landfills as well as at animal husbandry sites or pollutants which are introduced below ground (unlined latrines and burial sites) may contaminate the soil and travel down into aquifers causing degradation thereof. Pollutants of groundwater sources may include substances which occur as liquids (petroleum products) dissolved in water such as nitrates or are small enough to pass through the pores of soil such as bacteria. The movement of water within the aquifer consequently spreads these pollutants over a wide area, makes it unusable and spreads disease (DWA 2010b).

Even though groundwater can be deemed as an abundant resource, it still needs to be used sustainably to ensure continued recharge or replenishment. The maximum quantity of groundwater which could be developed in the country is estimated to be 6 billion m^3 per annum. Some groundwater resources take a long time to replenish and over-extraction of these sources will lead to depletion and in the case of coastal areas may allow salt water to replace it.

7.3 Overview of Challenges and Vulnerabilities: Current and Future

Current and future challenges or pressures on the country's water resources are and will be driven by its growing population, increased urbanisation as well as further developments which may include new mines and power stations, forestry and irrigation developments as well as pressures which arise due to poor water quality and management. These pressures or challenges are context-dependent and therefore occur within particular areas of the country.

Agriculture is the biggest water use sector in the country, using 62% of the country's available water resources mainly through irrigation (StatsSA 2010). Even though this is the biggest water user, the sector only contributes 2.5% to the country's GDP. Coal-fired power stations, nuclear stations as well as renewable power sources such as solar and wind power, also all need water to function and generate electricity. In terms of the mining sector, it only uses 8% of the country's total water resources however even though it accounts for a relatively small portion of the national water budget, it is a major water user in specific catchments where mining activities are concentrated and also adversely affects water quality. In terms of the commercial forestry sector, it also uses some of the country's national water budget but is regulated as a "stream-flow reduction activity" and mainly occurs in Mpumalanga and KwaZulu Natal Provinces (DWA 2010a).

The predominant challenges or vulnerabilities currently faced by South Africa are growing water quality problems and concerns, water losses as well as increased water stress and scarcity which are all driven by the constant growth in the human population, increased urbanisation or growing urban areas, further developments across the country and lastly poor or inefficient water planning and management. Future climate change has also been identified as a factor which will exacerbate these challenges and vulnerabilities if proper planning and management practices are not put in place. A brief discussion of these challenges or vulnerabilities, as well as climate change, now follow.

7.3.1 Growing Water Quality Problems

The degradation of water quality places further pressure on water demand as water becomes unfit for specific water uses due to being of unacceptable quality. Water quality is also a good indicator of socio-economic conditions as well as environmental awareness and attitude of its users. Changes in water quality can occur naturally along the length of a river but these changes are heavily influenced by human activities. These activities include agriculture, industries, mining and urban or rural settlements which produce nutrient concentrates (i.e. sewage effluent and fertilisers) as well as toxic substances (i.e. poisonous pollutants) which negatively affect water quality. The main causes of water pollution include the following:

- *Increased Urbanisation*: Affects water quality through physical disturbance of land, chemical pollution from activities, inadequate sewage collection and treatment as well as through increased nutrients from increased use of fertilisers.
- *Industries*: The wastewater produced by industrial activities affects the pH and colour of water, the amount of nutrients, temperature, amount of minerals and salts as well as turbidity.
- *Agriculture*: Agricultural activities increase soil erosion through physical disturbances and affect turbidity and amount of minerals and salts, the use of fertilisers and animal wastes/excreta contribute to increases in nutrients. Increase in pesticide use also contributes to water degradation.
- *Production and Use of Energy*: Increase in human population is accompanied with an increased energy requirement. Increased emissions from coal power stations result in increased emissions of sulphur and nitrogen oxides in the atmosphere which are the main causes of acid rain.
- *Deforestation*: The clearing of land which leads to increased soil erosion and increased sedimentation.
- *Destruction of Wetlands*: Leads to the destruction of habitats, destroys natural dams and ultimately removes natural filters.
- *Accidental Water Pollution*: This can arise from various activities or sources such as leaking or burst pipes and tanks as well as fires and oil spills. Damage caused can be of varying degrees depending on the quantity, toxicity and persistence of pollutants as well as the characteristics of the water body itself.

7.3 Overview of Challenges and Vulnerabilities: Current and Future

The main substances which are attributed to the degradation of water quality include possible poisonous chemicals (insecticides, heavy metals, problem products such as oil and petrol as well as chlorine and detergents) as well as fertilisers, sewage and blue-green algae which can be described as biological or microbial pollutants.

These main causes of water degradation and their associated substances consequently cause major water-quality problems across the country. These water quality challenges are predominantly induced by human activities (some problems are also due to natural causes). Industries producing chemical waste, mines introducing metals, wastewater treatment works discharging untreated or poorly treated effluents which introduce excessive nutrients, phosphates and coliforms as well as agriculture using pesticides, herbicides and fertilisers which introduces salts and other toxic substances into water bodies (Ashton 2009; DWA 2010c, 2011b; van der Merwe-Botha 2009). The main surface water quality challenges within the country include the following:

7.3.1.1 Salinisation

Salinity can be described as the quantity of total dissolved inorganic solids or salts whereby total dissolved solids (TDS) is used as an indicator to determine the state of the specific water resource. Agricultural return flows as well as urban and industrial runoff produce dissolved salts which consequently lead to increased salinity which is accompanied by a reduction in crop yields, scale formation, corrosion of water pipes as well as changes in freshwater biotic communities (DEAT 2006). High levels of salinity are deemed as a significant limiting factor in the fitness for use of water and is a persistent water quality problem throughout most of the country. Rivers can also be naturally saline due to geological conditions such as in the Northern, Western and Eastern Cape Provinces. Groundwater in some areas have also shown high levels of salinity above the recommended concentration for human use (Ashton 2009) however some aquatic ecosystems have naturally adapted to salinity levels.

7.3.1.2 Eutrophication

Eutrophication or the enrichment of water with nutrients (nitrates, phosphates and sewage) which encourages the growth of microscopic green plants and algae and promotes the growth of cyanobacteria which also presents a toxic threat to aquatic fauna and human users of water (DEAT 2006). The main consequence associated with this water quality problem includes the depletion of oxygen which leads to the mass mortalities of biota. Sources of nutrient pollution include domestic wastewater treatment plants, application of fertilisers as well as industrial and mining processes.

7.3.1.3 Micro-pollutants

Micro-pollutants tend to be highly localised and associated with specific industries or activities. It is associated with serious incidents of human health impacts for both humans and animals through uncontrolled exposure and has resulted in increased attention focussed on pollution through metals, carcinogens, synthetic chemicals, pharmaceuticals as well as veterinary and illicit drugs (Ashton 2009; Olujimi et al. 2010). Other ingredients in cosmetics, personal care products, as well as food supplements, have been found to possibly concentrate endocrine disrupting chemicals in the environment. Micro-pollutants can also enter water resources through accidental spills as well as via storm water runoff. Aquatic biodiversity is most at risk since the aquatic environment acts as a sink for hormonally active chemicals which may include industrial chemicals, pesticides, organochlorides, pharmaceuticals as well as natural and synthetic oestrogens and phytoestrogens (Olujimi et al. 2010; van der Merwe-Botha 2009).

7.3.1.4 Sedimentation

This water-quality problem is caused by runoff from land-based activities such as agriculture or poorly designed developments which carry sediment into rivers. Secondary effects of increased sediment load might include the decrease of useful lifespan of dams due to loss of storage capacity, decrease the lifespan of pumps and pipes and ultimately affects the integrity of rivers through sedimentation. It, therefore, can have substantial economic implications through infrastructure maintenance costs as well as increased costs in the management of water resources.

7.3.1.5 Micro-biological Pollutants

Micro-biological pollutants are mainly made up of bacteria which contaminates water which acts as a medium for the spread of water-related diseases which include dysentery, cholera, skin infections and typhoid. Most of these diseases are attributed to poor sanitation practices which arise from poorly maintained or lack of adequate infrastructure which is a widespread problem in South Africa (DEAT 2006; DWAF 2004).

The country's groundwater resources are also adversely affected by both pollution and over-abstraction in certain regions. Groundwater resource quality is poor and deteriorating over large parts of the country and can be attributed to various sources and sectors such as mining and industrial activities, effluent from municipal wastewater treatment facilities, storm-water runoff from urban especially informal settlements where sanitation treatment facilities are either inadequate or completely lacking, return flows from irrigated areas as well as effluent discharged from industries and other sources (DWA 2010b).

Groundwater is however predominantly classified as being of general good potable drinking water quality which needs little or no treatment on a large scale, however, there is an increase of salinity levels across the country which has resulted in deteriorating quality. Some groundwater resources have also been poorly managed mainly due to the lack of a structured approach to management and a general lack of knowledge and information about groundwater. Management is usually focussed on long-term sustainability of the resource in terms of quantity or yield and water quality is often neglected in many areas where groundwater is the sole source of water supply.

7.3.2 Water Losses and Increased Water Stress and Scarcity

The current water usage in South Africa is estimated between 15 and 16 billion m^3 per annum but can possibly increase to 20 billion m^3 per annum if effective metering and billing in urban and rural areas are not implemented. The supply and demand curve currently indicates that South Africa will face a supply–demand deficit of 17% by 2030. Current water usage is, therefore, exceeding reliable yield and has been clearly illustrated by the widespread water restrictions which have been put in place throughout the country with the recent (2015–2017) drought.

The ageing infrastructure, inadequate maintenance as well as repairs of existing infrastructure, slow responses to water leaks and bursts, shortcomings in technical competency in municipalities and lastly the water wastage culture has been a major challenge. These factors together with 95% of the country's resources being allocated in 2005 with increased pollution caused by mentioned sectors have further exacerbated the concerning nature of the country's water resources. The country will, therefore, be unable to sustain current patterns of water usage and discharge with these given factors together with current and anticipated population growth and socio-economic development (CSIR 2010; NEPAD 2013; DWS 2014b).

7.3.2.1 Widespread Water Losses

All water use sectors contribute to South Africa's high prevalence of water losses. The agricultural sector has been placed under increased pressure to maintain food security with the continued increase in the population which has, as a consequence, led to the intensification of farming practices and an increase in water abstraction. Increased water abstraction has in turn led to increased environmental problems such as lower groundwater and river flows, disappearance of wetlands due to alteration of drainage systems, extinction of some fauna and flora species caused by oxygen deficits, gradual salinization of water resources as well as increased leaching of nitrate, phosphate and pesticides which pollute both ground and surface waters (CSIR 2010; NEPAD 2013; GCIS 2015). Some water is also lost through unauthorised use for agricultural purposes and is included in the Non-Revenue Water (NRW) statistic of 39%.

Water losses are estimated at between 35 and 45% for irrigation schemes mostly due to falling into disrepair and some exceeding their economic lifespan. The agricultural sector primarily attributes to water losses through runoff losses, inefficient irrigation techniques and leakages, wasteful field application methods as well as cultivating thirsty crops which are not suited for the increased water-scarce environment. Other factors also include improper irrigation scheduling and soil type and soil preparation (Colvin 2015). Alien invasive species also play a role here where it is estimated that they consume 3 billion litres per annum of the country's water supply.

The mining and industry sector in the country is diverse and account for 16% of total water usage. The diversity of this sector causes the water use to be highly varied and obtain their water from Bulk water schemes or municipalities. This sector obtains its water supply by either abstracting water from a water resource regulated by the National Water Act or those services by water providers. Ill-maintained or inadequate water supply infrastructure mainly causes water losses through leakages. The pollution of surrounding water resources by wastewater as well as the dumping of waste materials is the other major factor which contributes to water losses as it causes water resources to become unfit for specific uses. The occurrence of acid mine drainage within the Gauteng, Mpumalanga Provinces and other mining areas is especially of high concern due to the large-scale damaging consequences it has had in all spheres.

NRW for municipal/domestic water use is estimated at 36.8% where 25.4% is due to physical leakages. Residential water use is just one component of the complete urban water use profile which also includes industrial, business or commercial, institutional and municipal water use as well as water loss (Hall and Watson 2000). The country's NRW statistic is almost the same as the world's average of 36.6% and does not compare well to other developed water-scarce countries where NRW is often less than 10%.

The low levels of payment in some parts of South Africa also have a major impact on NRW as there is little incentive to save water when a user has no intention of paying for it. The per capita use of water is also relatively high at 273 L per person per day which also emphasises the fact that the average citizen is still not completely aware of the magnitude of water scarcity in the country.

An estimated 91% of the country's population has access to basic services. Domestic water sector accounts for 30% of the total water requirement (25% urban and 5% rural areas). Water use, therefore, varies across income areas where high and medium income areas use more water than low-income areas. The water demand for domestic use is predicted to increase due to continued population growth, increased urbanisation as well as increase in the population having access to water services through addressing the backlog and expected improvement of living (CSIR 2010). Concerningly, rural municipalities have a NRW of 72.5% while metropolitan/urban municipalities an estimated 35.3% water loss. The high amount of water loss present at rural municipalities is attributed to poorer infrastructure and them having less resources to recover costs (KPMG 2014).

The key problem areas regarding South Africa's water losses have been identified as the following:

7.3 Overview of Challenges and Vulnerabilities: Current and Future

1. *Joint Responsibilities*: Most often the Technical and Financial departments in a water services division has certain responsibilities which in some cases can lead to issues in trying to establish the overall water balance of a municipality and associated estimated NRW. The continued replacement of councillors who generally have a lack of experience in water services as well as the importance of water conservation/water demand management (WC/WDM) also compounds this issue.
2. *Lack of Capacity*: there is an existing lack of capacity to monitor, regulate, enforce and support WC/WDM measures over the whole country. More capacity is required to facilitate this within all municipalities.
3. *Crisis Management*: Most municipalities are continually practicing crisis management with limited management information accompanied by poor decision-making processes, financial and technical management.
4. *Lack of Human Resources*: At an operational level there is an existing lack of human resources to perform basic functions which may include basic functions such as proactive maintenance, repair of leaks as well as the development of community awareness.
5. *Funding*: The funding of asset management, operation and maintenance as well as water loss/NRW reduction has not been prioritised even though most major municipalities or metropolitan areas have agreed that they can improve WC/WDM by prioritising their budgets. Some success stories do however exist.
6. *Lack of Proper Metering, Billing and Cost Recovery*: the finance function in most municipalities mostly requires major attention together with the training of councillors as well as financial and technical personnel.
7. *Overly Optimistic Water Savings of WDM Measures*: the time needed to achieve envisaged savings as well as associated costs are overly optimistic as water demand management is rarely a "quick fix". It needs to be implemented properly through a 5–10 year programme together with the continuous maintenance of interventions. Maintenance is rarely included in original project budgets but is essential for savings to be achieved and not be lost after project completion.
8. *Maintenance Perceived as a Problem*: Maintenance of WDM measures are perceived as problems rather than opportunities which can be used to create long-term employment opportunities in areas which often experience high levels of unemployment.
9. *Lack of Input*: More input is required from various other institutions to assist in helping with issues related to political support as well as high vacancies in some departments (DWS 2014b).

The enforcement of requirements of the Water Services Regulations can be one of the main key interventions which could be considered as it will assist in reducing water losses. Additional focus should, therefore, be placed on the following:

- Resolving intermittent supply by focussing on providing proper service;
- Addressing internal plumbing leakages by municipalities regardless if the services are being paid or not;
- Addressing visible leaks on distribution networks;

- Bulk metering and calculations of NRW;
- Accurate metering, reading and billing; and
- Creating consumer awareness and eliminate inefficient use.

7.3.2.2 Increased Water Stress and Scarcity

Water scarcity has increasingly been placed under the spotlight and high on some agendas especially during the last decade. Water scarcity is mainly driven by growing populations, increased urbanisation as well as extreme weather events driven by climate change. Political apathy has also been labelled as a driver for the looming water crisis especially when it exacerbated the effects of the last severe drought (2015–2017) mainly due to massive water losses caused by inadequate maintenance of ageing infrastructure, poor management of water distribution networks and ultimately the lack of proper response to warnings given by scientific studies since 2000.

Water scarcity has become the most severe in the Western Cape province as well as in certain parts of the Northern and Eastern Cape provinces. The current water crisis experienced in Cape Town has become severe. Major dams supplying the city averaged at approximately 36% in 2017, whereas they were 62% in 2016. A usage limit of 87 L per person has been implemented to attempt to decrease water usage within the city.

As indicated previously, climate change has become a significant influence on current and future water stress and scarcity within the country. Climate change affects both temperature as well as rainfall. Temperature is predicted to increase and rainfall decrease as well as become more erratic. Current estimates are that South Africa will experience a reduction of 10% of its average rainfall causing surface runoff to decrease up to 50–75% by 2025. Climate change has already attributed to the growing water crisis in the country especially within the last decade as rainfall has become more and more infrequent.

South Africa is considered to be very vulnerable to climate change as its water resources are not evenly distributed. Increased limited water availability in certain regions will cause people to have to migrate to other areas with greater water availability which will consequently place more pressure on major water resources. The increase of natural disasters or extreme weather events such as droughts can also increase the number of destitute people migrating to urban areas which in turn exacerbates pressures on natural resources and increases pollution and the occurrence of disease (Zietsman 2011).

During the recent drought, the first response in South Africa was to drill boreholes and make use of groundwater. Groundwater can, therefore, be considered as a response to relieve water stress within some regions as it experiences much lower evaporation and declines slower during drought periods. However, it should be noted that since rainfall mainly recharges aquifers, long-term climate change may have dramatic effects on groundwater resources as well. Groundwater, therefore, needs to be managed sustainably to ensure that they do not become stressed.

7.3 Overview of Challenges and Vulnerabilities: Current and Future

The further degradation of water resources through the various previously described avenues will also be exacerbated by climate change in terms of frequency and duration. Nutrient loading or eutrophication will especially be exacerbated. Increased nutrient loading, as well as cyanobacteria blooms, will result in progressive outbreaks in health-associated risks which will negatively impact all people in South Africa. Rural communities will be the most impacted as they rely on ecosystem goods and services as a direct lifeline for natural resources in terms of harvesting, agricultural practices, consumption as well as other livelihood strategies.

For the country to improve on its freshwater resources, it needs to improve on its overall management approaches and treatment technologies (Oberholster and Ashton 2008; King et al. 2009; Pitman 2011). The calculations for the country's future water consumption do not take account of all the water used by the informal farming sector and rural communities, nor the water that is needed for shared water courses and ecosystem maintenance (SADC 1995). These amounts could increase the volumes by 20–30% above the original estimates (Pallet 1997; Kundzewicz and Krysanova 2010).

If these estimates are valid, the deduction can be made that the effective limit of South Africa's exploitable resources will be reached within 15–30 years, as estimated in 1997 (Basson et al. 1997; Pallet 1997; Ashton 1999; Kundzewicz and Krysanova 2010). The highest and lowest projections, with existing and new technologies, for water demand in the country as against available surface water, groundwater and recycled water.

Thus, the South African Department of Water and Sanitation (DWS) have developed several guidelines in order to ensure that, depending on water usage, the quality of the water is of the correct standard. Water quality is negatively affected by numerous anthropogenic activities, and by either point or diffuse source pollutants. South Africa's freshwater resources are fully allocated with the result that the water quality of these resources needs to be effectively managed and maintained.

To avert a predicament, water demand needs to be managed in conjunction with water conservation to ensure that all sectors improve their water use efficiency for the development of potential water savings. Water demand needs to focus on reducing water losses, increasing water productivity and re-allocating water resources but also take water pollution into account. Focus, therefore, needs to be placed on minimising non-productive losses from water systems and preventing degradation of water sources.

References

Arntzen J (2001) Sustainable water management in southern Africa: an integrated perspective. In: Gash JHC, Odada EO, Oyebande L, Schulze RE (eds) Freshwater resources in Africa. Proceedings of a workshop, Nairobi, Kenya, Oct 1999, pp 81–87. Biospheric Aspects of the Hydrological Cycle

Ashton PJ (1999) Integrated catchment management balancing resource conservation with utilization. In: Proceedings of the South African water management conference, Kempton Park, South Africa, 2–4 Aug, 12 pp

Ashton PJ (2009) An overview of the current status of water quality in South Africa and possible future trends of change. Water Ecosystems and Human Health Research Group, Natural Resources and the Environment Unit, CSIR, Pretoria

Ashton PJ, Hardwick D, Breen C (2008) Changes in water availability and demand within South Africa's shared river basins as determinants or regional, social and ecological resilience. In: Burns M, Weaver A (eds) Exploring sustainability science: a southern African perspective. African Sun Media, Stellenbosch, South Africa

Basson MS, van Niekerk PH, van Rooyen JN (1997) Overview of water resources availability and utilisation in South Africa. Department of Water and Sanitation and BKS (Pty) Ltd, Pretoria

Beekman HE, Saayman I, Hughes S (2003) Vulnerability of water resources to environmental change in southern Africa. A report for the Pan African START Secretariat and UNEP

Bere T (2007) The assessment of nutrient loading and retention in the upper segment of the Chinyika River, Harare: implications for eutrophication control. Water SA 33:279–284

Biggs R, Simons H, Bakkenes M, Scholes RJ, Eickhout B, van Vuuren D, Alkemade R (2008) Scenarios of biodiversity loss in southern Africa in the 21st century. Global Environ Change 18:296–309

Binns T, Illgner P, Nel E (2001) Water shortage, deforestation and development: South Africa's 'Working for Water' programme. Land Degrad Dev 12:341–355

Braune E, Xu Y (2008) Groundwater management issues in southern Africa—an IWRM perspective. Water SA 34:699–706

Christelis G, Hunger G, Mulele O, Mangisi N, Mannathoko I, van Wyk E, Braune E, Heyns P (2010) Towards transboundary aquifer management in southern Africa. Available via http://www.siagua.org/sites/default/files/documentos/documentos/sur_africa.pdf. Accessed on 26 Feb 2018

Cobbing JE, Hobbs PJ, Meyer R, Davies J (2008) A critical overview of transboundary aquifers shared by South Africa. Hydrogeol J 16:1207–1214

Colvin C (2015) Curbing water loss in agriculture: Q&A with Christine Colvin. Available via http://www.bizcommunity.com/Article/196/643/135179.html. Accessed on 26 Feb 2018

Conca K (2006) Governing water: contentious transnational politics and global institution building. MIT Press, Cambridge

CSIR (Council for Scientific and Industrial Research) (2010) A CSIR perspective on water in South Africa. CSIR report no. CSIR/NRE/PW/IR/2011/0012/A

DEAT (Department of Environmental Affairs and Tourism) (2006) South Africa environment outlook. A report on the state of the environment. Department of Environmental Affairs and Tourism, Pretoria

Descheemaeker K, Amede T, Haileslassie A (2010) Improving water productivity in mixed crop-livestock farming systems of sub-Saharan Africa. Agric Water Manage 97:579–586

DWA (Department of Water Affairs) (2010a) Assessment of the ultimate potential and future marginal costs of water resources in South Africa. Project Completed by BKS, Pretoria

DWA (Department of Water Affairs) (2010b) Groundwater strategy 2010. Water resources planning systems. Department of Water Affairs, Pretoria

DWA (Department of Water Affairs) (2010c) National desalination strategy, final revision for approval, Oct 2010. Department of Water Affairs, Pretoria

DWA (Department of Water Affairs) (2011a) Integrated water resource planning for South Africa: a situation analysis. Department of Water Affairs, Pretoria

DWA (Department of Water Affairs) (2011b) Planning level review of water quality in South Africa. Resource directed management of water quality. Department of Water Affairs, Pretoria

DWAF (Department of Water Affairs and Forestry) (2004) National water resource strategy, 1st edn, Sept 2004. Department of Water Affairs and Forestry, Pretoria

DWS (Department of Water and Sanitation) (2014a) South Africa yearbook 2013/2014: water affairs. Available via http://www.southafrica-newyork.net/consulate/Yearbook_2014/2013-4Water_Affairs.pdf. Accessed on 26 Feb 2018

DWS (Department of Water and Sanitation) (2014b) National water resource strategy, 2nd edn. Government Gazette, Pretoria

References

Edwards AC, Withers PJA (2008) Transport and delivery of suspended solids, nitrogen and phosphorus from various sources to freshwaters in the UK. J Hydrol 350:144–153

FAO (Food and Agriculture Organisation) (2006) Aquastat land and water use database. FAO, Rome. Available via http://www.fao.org/ag/agl/aglw/aquastat/main/index.stm Accessed on 26 Feb 2018

GCIS (Government Communication and Information System) (2015) Agriculture. Available via http://www.gcis.gov.za/sites/www.gcis.gov.za/files/docs/resourcecentre/Agriculture2015.pdf. Accessed on 26 Feb 2018

Gleditsch NP, Furlong K, Hegre H, Lacina B, Owen T (2004) Conflicts over shared rivers: resource scarcity or fuzzy boundaries? Paper given at the 45th annual convention of the international studies association on 'hegemony and its discontents', Montreal, Canada, 17–20 Mar 2004

Grey D, Sadoff CW (2007) Sink or swim? Water security for growth and development. Water Policy 9:545–571

Hall E, Watson M (2000) Urban water consumption. In: IWA, managing water and waste in the new millennium, Midrand, Johannesburg, South Africa, 23–26 May 2000, pp 1–9

Hirji R, Johnson P, Maro P, Matiza Chiuta T (eds) (2002) Defining and mainstreaming environmental sustainability in water resources management in southern Africa. SADC, IUCN, SARDC, World Bank, Maseru/Harare/Washington, DC

King NA, Maree G, Muir A (2009) In: Strydom HA, King ND (eds) Environmental management in South Africa: chapter 13: freshwater systems, 2nd edn. Juta

KPMG (2014) SA needs to curb water wastage. KPMG South Africa Blog. Available via http://www.sablog.kpmg.co.za/2014/04/saneedscurbwaterwastage/. Accessed on 26 Feb 2018

Kundzewicz ZW, Krysanova V (2010) Climate change and stream water quality in the multi-factor context. Clim Change 103:353–362

Molapo P, Pandey SK, Puyoo S (2000) Groundwater resource management in the SADC region: a field of regional cooperation. In: IAH 2000 conference, Cape Town

NEPAD (The New Partnership for Africa's Development) (2013) Strategic overview for the water sector in South Africa. DWAF. Available via http://nepadwatercoe.org/wp-content/uploads/Strategic-Overview-of-the-Water-Sector-in-South-Africa-2013.pdf

Ngcobo S, Jewitt GPW, Stuart-Hill SI, Warburton ML (2013) Impacts of global change on southern African water resources systems. Environ Sustain 5:655–666

Nicholson SE (2000) The nature of rainfall variability over Africa on time scales of decades to millenia. Global Planet Change 26:137–158

Oberholster PJ, Ashton PJ (2008) An overview of the current status of water quality and eutrophication in South African rivers and reservoirs: state of the nation report. Parliamentary Grant Deliverable

Olujimi OO, Fatoki OS, Odendaal JP, Okonkwo JO (2010) Endocrine disrupting chemicals (phenol and phthalates) in the South African environment: a need for more monitoring. Water SA 36. Pretoria

Pallet J (1997) Sharing water in southern Africa. Desert Research Foundation of Namibia, Windhoek, Namibia

Parnell S, Walawege R (2011) Sub-Saharan African urbanisation and global environmental change. Global Environ Change 21:12–20

Pitman WV (2011) Overview of water resource assessment in South Africa: current state and future challenges. Water SA 37:659–664

SADC (Southern African Development Community) (1995) Protocol on the shared watercourse systems in the Southern African Development Community Region. SADC Council of Ministers, Gaborone, Botswana

SADC (Southern African Development Community) (2005) Regional strategic action plan for integrated water resource management

Schulze RE (ed) (2005) Climate change and water resources in southern Africa: studies on scenarios, impacts, vulnerabilities and adaptation. Water Research Commission, Pretoria, RSA

StatsSA (Statistics South Africa) (2010) Water management areas in South Africa. Discussion document: D0405.8. Statistics South Africa, Pretoria

Turton AR (2003) Environmental security: a southern African perspective on transboundary water resource management. Environ Change Secur Proj Rep 9:75–87

Turton AR (2004) The evolution of water management institutions in select southern African international river basins. In: Tortajada C, Unver O, Biswas AK (eds) Water as a focus for regional development. Oxford University Press, London

Turton AR (2005) A critical assessment of the river basins at risk in the southern African hydropolitical complex. Paper given at a workshop hosted by the Third World Centre for Water Management and the Helsinki University of Technology on 'The Management of International Rivers and Lakes', Helsinki, Finland, 17–19 Aug 2005

Turton A (2008) The state of water resources in southern Africa: what the beverage industry needs to know. CSIR technical report

Turton AR, Earle A (2005) Post-apartheid institutional development in selected southern African international river basins. In: Gopalakrishnan C, Tortajada C, Biswas AK (eds) Water institutions: policies, performance and prospects. Springer-Verlag, Berlin

Turton AR, Meissner R, Mampane PM, Seremo O (2004) A hydropolitical history of South Africa's international river basins. Water research commission report no. 1220/1/04. WRC, Pretoria

Turton A, Patrick M, Cobbing J, Julien F (2006) The challenges of groundwater in southern Africa. Navigating Peace, Wilson Centre, Aug 2006, no. 2

UNESCO (United Nations Educational, Scientific and Cultural Organisation) (2011) United Nations Educational, Scientific and Cultural Organisation

van der Merwe-Botha M (2009) Water quality: a vital dimension of water security. Development planning division. Working paper series no. 14. DBSA, Midrand

WCD (World Commission on Dams) (2000) Dams and development: a new framework for decision-making. Earthscan, London

WWF (World Wildlife Fund) (2013) An introduction to South Africa's water source areas. WWF-SA

Zietsman L (ed) (2011) Observations on environmental change in South Africa. South African Environmental Observation Network (SAEON), Sun Media, Stellenbosch

Chapter 8
Establishing South Africa's Current Water Quality Risk Areas

South Africa developed its National Water Policy underpinned by integrated water resource management. International good practice was recommended and followed which resulted in the decentralisation of water management and the establishment of water management institutions which are based on hydrological instead of political boundaries. Nineteen Water Management Areas (WMA) were established with envisaged establishment of a Catchment Management Agency (CMA) for each. This was subsequently decreased to nine as concerns were raised regarding the capacity of the country to manage and support 19 CMAs.

For the purpose of this book, water quality data of selected physical, chemical and biological water quality parameters were obtained from Department of Water and Sanitation: Resource Quality Services Department which were consequently structured, validated and analysed. Four risk categories were developed to classify and establish water quality risk areas for each WMA.

8.1 Brief Background to South Africa's Water Management Areas

South Africa developed its National Water Policy for South Africa (1997) and the National Water Act (Act 36 of 1998) through extensive public participation as well as international expertise and advice, underpinned by integrated water resource management. This resulted in the recommendation to follow international good practice through the decentralisation of water management and the establishment of water management institutions which are based on hydrological instead of political boundaries. The country proceeded to develop the National Water Resource Strategy in 2004 which led to the development of 19 Water Management Areas (WMAs) as well as the envisaged establishment of a Catchment Management Agency (CMA) for each (DWS 2016). Concerns were raised regarding the capacity of the country to manage and support 19 CMAs and a decision was consequently made to reduce the number to nine (Fig. 8.1).

The development of CMAs was drawn from international experience which identifies multiple key drivers for catchment-based management of water resources. The main needs identified for the establishment of CMAs included:

- Achieving integrated management of a catchment;
- Facilitating stakeholder participation in decision-making and management of water resources; and
- Separating policy and national strategy functions of the Ministry/Department and the operational functions of the CMA.

It should be noted that these WMA boundaries do not correlate with South Africa's administrative boundaries. Factors that were taken into consideration were institutional efficiency, the financial self-sufficiency of the water consumers, the centres or locations of economic activity, social development patterns, the distribution of water resource infrastructures, and lastly, the centres of water-related expertise for future assistance (DWAF 2004).

The DWS has however recently announced that it is aiming to dissolve or merge the nine CMAs into a single CMA however this is still being considered and under discussion.

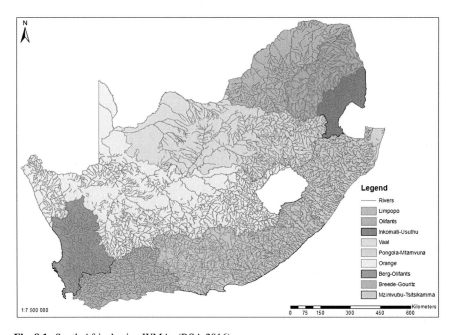

Fig. 8.1 South Africa's nine WMAs (RSA 2016)

8.2 Establishing South Africa's Water Quality Risk Areas

For the purpose of this book, water quality data were collected for the nine WMAs which were established in 2016. Water quality data were collected for the period of July 2011–June 2017 from the DWS: Resource Quality Services. Water quality data were still grouped under the previous 19 WMAs and were consequently consolidated and grouped into the nine WMAs. The types of water quality sampling stations included

- Dams/barrages;
- Rivers;
- Springs/eyes;
- Wetlands;
- Estuaries/lagoons; and
- Wastewater treatment works (WWTWs).

Collected water quality data were structured and validated. The water quality data obtained from the water quality sampling points were measured monthly, weekly and, in some cases, daily throughout the year, but at no scheduled time and on no fixed day or week. An overall average for the mentioned period was calculated for each of the water quality parameters used at each station.

Some sampling stations were excluded due to inadequate data recordings. Sampling stations were therefore excluded in the case of the station having recorded less than four measurements within a year. Some water quality sampling stations were also not included in this research due the station having measured less than four parameters with less than four recorded measurements within a year. The "Four by Four" (4×4) rule was therefore followed as recommended by the Canadian Council of Ministers for the Environment. This rule ensures that only sampling stations that regularly monitor the relevant parameter are included and eliminates stations that have only three monitoring phases per year. The evaluation was therefore limited to these sampling stations which had adequate replication in an attempt to ensure high quality data and representation.

A total of 2,438 water quality sample stations were included in the evaluation in terms of physical and chemical water quality parameters. Only 318 water sample stations measured *Chlorophyll a* and 1,619 water quality sample stations measured *Faecal coliform* levels (Figs. 8.2, 8.3 and 8.4).

In terms of water quality parameters, there is no single parameter that can be used to describe the overall water quality for any water body. A wide variety of water quality parameters can and should be used to obtain a holistic and accurate view of a water body's water quality in terms of environmental and human health (UNEP 2007). Water quality parameters were selected according to the following rules:

- The water quality parameter needs to have available national water quality indices or guidelines.
- The water quality parameter must have been commonly measured and reported by the DWS water quality sampling stations.

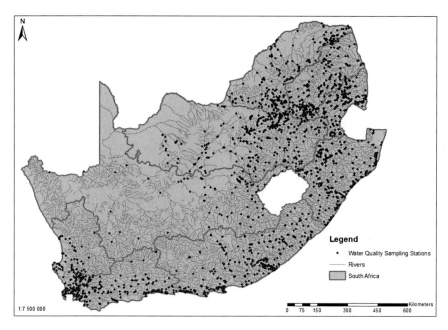

Fig. 8.2 Evaluated physical and chemical water quality parameter sample stations

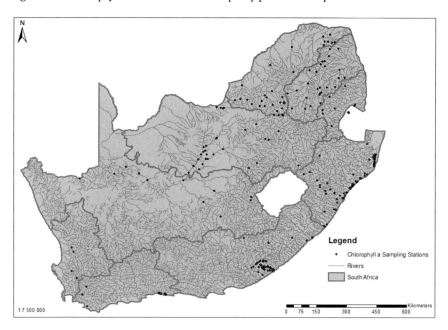

Fig. 8.3 Evaluated *Chlorophyll a* sample stations

8.2 Establishing South Africa's Water Quality Risk Areas

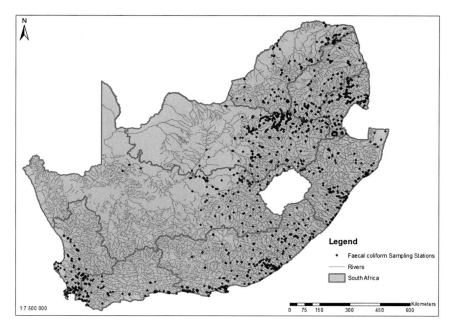

Fig. 8.4 Evaluated *Faecal coliform* sample stations

- The water quality parameter needs to have a representation percentage of a minimum of 80% (50% in the case of biological parameters).
- Water quality parameters which are characterised by the occurrence or measurement of non-detectable values needs to be excluded due to the possibility of bias.

South Africa's National DWS water quality guidelines for the following water uses were used for the selected water quality parameters:

- Domestic use,
- Aquatic Ecosystems,
- Irrigation, and
- Industrial use.

Subsequently a total of 11 water quality parameters[1] were selected included the following:

- *Physical parameters*: pH and Electrical Conductivity (EC);
- *Chemical parameters*: Calcium, Chloride, Sodium, Ammonia (NH_4), Nitrates (NO_3), Phosphate (PO_4) and Sulphate (SO_4); and
- *Biological/Microbiological parameters*: *Chlorophyll a* and *Faecal coliform*.

All of the selected water quality parameters had relevant domestic use water quality guidelines. In terms of the other water use water quality guidelines, not all

---

[1] For reference, the water quality guidelines and standards for each of the identified water quality parameters are listed in Appendix.

of the selected water quality parameters had a guideline. Only chloride, ammonia, nitrate and phosphate had applicable water quality guidelines in terms of Aquatic ecosystems. pH, EC, chloride, sodium, ammonia and nitrates had applicable water quality guidelines for irrigation use and lastly pH, EC, chloride and sulphate had applicable water quality guidelines in terms of industrial use.

In terms of Microbiological/Biological water quality parameters the following water quality standards were used as not all of the mentioned water uses have guidelines. Domestic use and aquatic ecosystems water quality guidelines were used for *Chlorophyll a*, and domestic use and irrigation water quality guidelines were used for *Faecal coliform*.

To ultimately establish water quality risk areas, for the purpose of this book, four risk categories were developed. These risk categories included the following:

- 0—No-Risk Area (All water quality parameters are of ideal to acceptable standard);
- 1—Low-Risk Area (One or two water quality parameters are of tolerable to unacceptable standard);
- 2—Medium-Risk Area (Potential Future Risk Area—50% of water quality parameters are of tolerable to unacceptable standard); and
- 3—High-Risk Area (Majority >80% of water quality parameters are of tolerable to unacceptable standard).

The most up to date land cover dataset (2013/2014) were also obtained from South African National Biodiversity Institute (SANBI) to assist in establishing main causes for the established water quality risk areas and possible consequences thereof.

To ultimately achieve the establishment of South Africa's water quality risk areas, the research procedure was differentiated into three phases. These phases, with their main steps or actions, are presented in Fig. 8.5.

Some shortcomings in the water quality data were identified and included the following. Some water sampling stations contained some gaps and inconsistencies in the form of missing values or extreme values. Linear interpolation was used to reprocess the data in order to fill in some of the missing parts in the data. Table 8.1 presents the important sources of uncertainty concerning water quality data. Not all of these uncertainties as listed in table were applicable or relevant. Instrument errors were found to have been relevant as extreme values were present. These measurements were consequently identified as extreme values, were removed and replaced by an interpolated value. Furthermore, the variation in sampling frequency was also identified as an uncertainty as the sampling frequency differed between some water quality sampling stations. This uncertainty was addressed by structuring the data in a uniform way through the calculation of an overall average.

Some missing values were present in this data range. These gaps or missing values could be attributed to the following. First, the instruments used for measuring the quality of the water could break down or become faulty and could therefore register inaccurate measurements. It proved to be a lengthy process to repair the instruments as in most cases parts needed to be ordered from overseas. Much time passed before these parts arrived in the country and could be installed.

8.2 Establishing South Africa's Water Quality Risk Areas

PHASE 1: DATA COLLECTION, STRUCTURING AND CLEANING

Step 1: *Collect* data.

Step 2: *Clean* data: Identify extreme/ inaccurate/ missing values and apply interpolation techniques where needed.

Step 3: *Structure* data into databases (WMAs): Calculate overall average for each water quality sampling station and parameter to make data suitable for analysis.

PHASE 2: DATA ANALYSIS

Step 1: *Evaluate water quality*: All water quality sampling stations and their water quality parameters were evaluated according to the National water quality guidelines in terms of the mentioned water uses. Water quality parameters which were measured to be of tolerable or unacceptable standard were identified for each station according to the specific water use standard.

Step 2: *Develop water quality risk categories*: Developed four categories to illustrate water quality risk for each station.

Step 3: *Establish water quality risk:* Water quality risk was established for each water quality sampling station in terms of each mentioned water use standard.

PHASE 3: INTERPRETATION OF RESULTS

Step 1: *Evaluate temporal and spatial trends*: Identify important and significant variations in the selected water quality parameters.

Step 2: *Evaluate relationships*: Identify significant relationships between established water quality risk areas and land cover.

Step 3: *Evaluate scenarios:* Identify areas of concern in terms of environmental, social and economic impacts.

Fig. 8.5 Phases in the research procedure and their main actions

As a result of the complexities of the data, a heuristic approach was used in addition to the above-mentioned statistical approaches, with the main focus being on assessing the outcome of the linear interpolation, which had been completed. This approach was also necessary for data reflecting unrealistic values. An example illustrating the application of incorrect data can be quoted in the case of pH values of 17 and 18 that were measured at some water quality sampling points, all of which were clearly incorrect. Interpolation was required in such cases and consequently carried out.

Table 8.1 Most important uncertainties in water quality data (van Loon et al. 2005; Rode and Suhr 2007)

Field instruments	Sampling location	Representative sampling	Laboratory analysis	Load calculation
Instrument errors	Mixing of large tributaries	High spatial variation within cross section	Sampling conservation	Sampling frequency
Instrument calibration errors	Point source inputs	High temporal variation (due to point source inputs and flood events)	Sample transport	Sampling period
	Impoundments and dead zones	Sampling volume	Instrument errors	Choice of extrapolation method
		Sampling duration	Laboratory-induced uncertainties	

Lastly, the method used in the collection and chemical analysis of the water quality samples could also be regarded as a shortcoming. According to Davies and Day (1998), this type of chemical analysis is unsatisfactory as, although it provides accurate measurements in terms of the quantity of the selected individual substances present in the water body, it only takes the water that flows past a specific point into account at the time of collection. Thus, although these water quality samples were collected throughout a 24-h period in order to calculate the average monthly readings for the specific water quality parameter, the data were still only collected once a month and represented by a single average value (Davies and Day 1998).

Davies and Day (1998) concluded that such readings may be subjective to a particular degree of inaccuracy as variations within these measured concentrations of the selected water quality parameter can fluctuate significantly over the period of a month.

Shortcomings were also identified regarding the use of national land cover data. These included possible inaccuracies. A logical error test was consequently completed through the completion of a sensitivity analysis to establish whether significant errors were present. Very few errors were, however, identified through this testing process. Errors were corrected when necessary. Mixed classes or classification structure is also a recognised limitation of land use data. Mixed classes or classification structure was not a limitation due to the use of a hierarchical classification design namely the "Standard Land Cover Classification Scheme for Remote Sensing Applications in South Africa" (Burrough 1990; Liou et al. 2004; Longley et al. 2005).

Lastly, the land cover dataset is but a single "snapshot" of a period in time. Owing to the dynamic nature of the land cover and land use, this issue also needs to be taken into account when identifying patterns and making conclusions.

The nine WMAs have been divided into regions and three chapters. Chapter 9, the northern region which includes the Limpopo, Olifants and the Inkomati-Usuthu WMAs. Chapter 10, the central region, which includes the Vaal, Pongola-Mtamvuna and Orange WMAs and lastly Chap. 11, the southern region, which includes the Berg-Olifants, Breede-Gouritz and Mzimvubu-Tsitsikamma WMAs.

These chapters will give a brief overview for each WMA in terms of its main characteristics as well as the location and amount of water quality sampling points. This will be followed by the evaluation of water quality risk areas based on the mentioned national water quality guidelines, i.e. domestic use, aquatic ecosystems, irrigation as well as industrial use in terms of the selected physical and chemical water quality parameters. Each WMA will also be evaluated in terms of establishing main risk areas for *Chlorophyll a* according to domestic use and aquatic ecosystem water quality standards and *Faecal coliform* according to domestic use and irrigation water quality standards. The main water quality issues will also be discussed for each of the WMAs. The results for each of the nine WMAs now follow in the following three chapters through detailed discussions and evaluations.

References

Burrough PA (1990) Principles of geographical information systems for land resource assessment. Clarendon Press, Oxford

Davies B, Day JA (1998) Vanishing waters. University of Cape Town Press, Cape Town

DWAF (Department of Water Affairs and Forestry) (2004) Department of Water Affairs and Forestry, South Africa. Report no. PC 000/00/22602. Annual operating analysis for the total integrated Vaal River system (2002/2003). Compiled by WRP Consulting Engineers

DWS (Department of Water and Sanitation) (2016) The water management areas starter pack. Directorate: Catchment Management

Liou SM, Lo SL, Wang SH (2004) A generalised water quality index for Taiwan. Environ Monit Assess 96:35–52

Longley PA, Goodchild MF, Maguire DJ, Rhind DW (2005) Geographic information systems and science: chapter 6, 2nd edn. Wiley, Hoboken, NJ

Rode M, Suhr U (2007) Uncertainties in selected river water quality data. Hydrol Earth Syst Sci 11:863–874

RSA (Republic of South Africa) (2016) Government Gazette No. 40279, 16 Sept 2016. Pretoria, South Arica

UNEP (United Nations Environmental Programme) (2007) Global drinking water quality index development and sensitivity analysis report. Available via http://www.gemswater.org

van Loon E, Brown J, Heuvelink G (2005) Guidelines for assessing data uncertainty in river basin management studies: methodological considerations. In: van Loon E, Refsgaard JC (eds) HarmoniRiB report

Chapter 9
Current Water Quality Risk Areas for Limpopo, Olifants and the Inkomati-Usuthu WMAs

Focus is placed on the northern WMAs of South Africa which include the Limpopo, Olifants as well as the Inkomati-Usuthu WMAs. The Limpopo, Olifants and Inkomati-Usuthu WMAs were all found to be predominantly of low risk in terms of the selected physical and chemical water quality parameters but of concerningly high risk in terms of *Chlorophyll a* and *Faecal coliform*.

Significant risk areas were, however, established for all of the WMAs and directly correlate with the extent of modification of water sources or areas. Significant risk areas were predominantly established downstream or within close proximity of urban centres, cultivated areas, mining developments as well as WWTWs. WWTWs are of great concern for the whole northern region as most sampling stations recorded tolerable to unacceptable standards of most or all selected water quality parameters especially in terms of *Faecal coliform*. Most of the WWTWs facilities within these WMAs do not comply with set standards and can be attributed to these facilities being mismanaged, inadequate or in need of proper maintenance. This needs to be addressed to avoid future significant environmental human health problems and risks.

9.1 Limpopo WMA

9.1.1 WMA Overview

The Limpopo WMA comprises of the Crocodile West and Marico, Limpopo as well as the Luvuvhu catchment areas and is predominantly characterised by low rainfall and significant inter-dependencies for water resources between catchments and neighbouring WMAs. The major rivers within the Limpopo catchment are the Matlabas, Mokolo, Lephalala, Mogalakwena, Sand, Nzhelele and Nwanedi. Few sites are available for the construction of major dams mainly due to the flatness of the WMAs terrain as well as the aridity. Surface water potential has also largely been developed. Groundwater is used extensively due to relatively favourable formations and is overexploited in certain areas. Several inter-water management area transfers exist which brings additional water into the WMA (Fig. 9.1).

The WMA is mainly centred around game, livestock and irrigation farming with increasing mining developments. A transfer from the Crocodile West catchment to the Mokolo catchment is planned to support the increase in mining and power generation in the Lephalale area.

The Luvuvhu River sub-catchment is located in the north-east region of the WMA, and is the only well-watered catchment in the WMA with the Mutale River being its main tributary. Thohoyandou is the main urban area with large rural populations scattered across the area. Groundwater is utilised on a large scale by all water use sectors. The Limpopo catchment includes the Matlabas, Mokolo, Lephalala, Mogalakwena, Sand and Nzhelele rivers which together with smaller tributaries flow northwards towards the Limpopo River. The catchment varies from being highly developed urban areas such as Polokwane (50% of urban population), Lephalale and Mokopane to rural communities relying on subsistence farming. The catchment is also characterised by irrigation areas as well as mining in the form of large coal and platinum mining operations located close to Lephalale.

The Crocodile West and Marico catchment, located in the south-east, supports major economic activities and an urban population of approximately 5 million. It is consequently the second most populated catchment in the country with the largest proportional contribution to the country's national economy. The catchment is highly

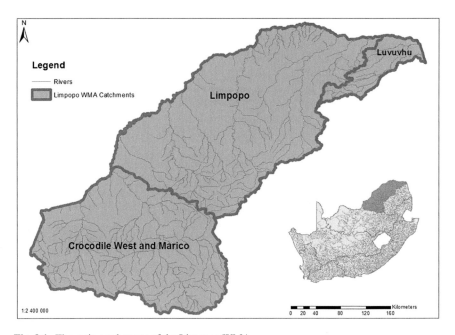

Fig. 9.1 The main catchments of the Limpopo WMA

9.1 Limpopo WMA

altered by catchment development (dominated by urban areas and industrial complexes), extensive irrigation along major rivers with game and livestock farming occurring in other parts.

The development and utilisation of surface water occurring naturally have reached its full potential in the WMA. Increasing quantities of effluent return flow from urban and industrial areas offer considerable potential for reuse, but the effluent is at the same time a major cause of pollution in some rivers. Population and economic growth centred on the Johannesburg–Pretoria metropolitan complex as well as mining developments are expected to continue strongly as well.

Water quality monitoring of surface water resources are limited, with large parts of the WMA not having any monitoring data available especially in the central and north-eastern parts of the Limpopo catchment (Fig. 9.2). A total of 264 sampling stations had suitable data for the time period and were evaluated. Of these 264 sampling stations, 195 are river-, 37 are dams/barrages-, 11 are spring/eyes- and 21 wastewater treatment works (WWTWs) sampling points. River and dam/barrage sampling points are distributed across the WMA while spring/eyes are located mostly on the outskirts of the WMA. WWTWs are predominantly located closer to urban built-up areas and in some cases mining developments.

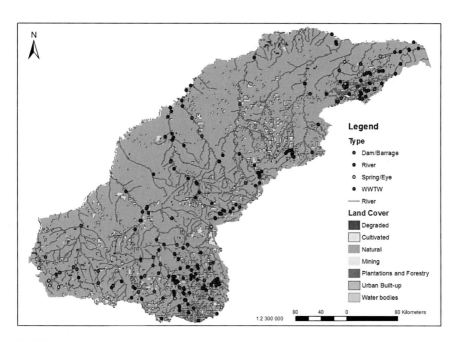

Fig. 9.2 Water quality sampling sites used and land cover of the Limpopo WMA

9.1.2 Risk Areas for Domestic Use

The Limpopo WMA is mostly dominated by no- and low-risk areas especially in the less developed regions of the WMA. A total of 189 (71.5%) of sampling stations were found to be of no risk and 53 (20%) of sampling stations low-risk areas (Fig. 9.3).

Most of the significant risk areas (Risk level 2 and 3) are in close proximity to urban built-up areas and in some cases cultivated areas and mining developments. Medium-risk areas include six river sampling points and seven WWTWs located in the Crocodile West and Marico catchment, located in the south-east of the WMA, and two WWTWs sample points in the Limpopo sub-catchment. These medium risk areas are predominantly located closely downstream from urban built-up, cultivated and mining land cover areas. High-risk areas were recorded at 6 river- and one WWTW sampling points, all located in the Crocodile West and Marico catchment. These sampling stations had tolerable to unacceptable levels of all or most of the selected physical and chemical water quality parameters and are of major concern. These areas are once again located directly downstream from urban built-up, cultivated and mining developments. Water at these locations can therefore not be directly used for domestic use.

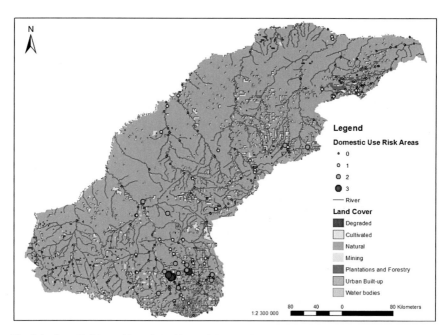

Fig. 9.3 Overall risk profile and significant risk areas for the Limpopo WMA (domestic use standards)

9.1.3 Risk Areas for Aquatic Ecosystems

In terms of aquatic ecosystem water quality standards, no sampling stations qualified as having no risk. All sampling stations varied from low, medium to high risk. The WMA is dominated by low risk, however, 9% of sampling stations are of medium risk and 15% high (Fig. 9.4). Therefore, in terms of aquatic health, the WMA is predominantly degraded at varying degrees.

Most of the significant risk areas (Risk level 2 and 3) are in close proximity to urban built-up areas as well as cultivated areas and mining developments. Medium risk areas include river, dam/barrage and WWTWs sampling points located across the WMA. These medium risk areas are predominantly located closely downstream from urban built-up, cultivated and mining land cover areas. High-risk areas were recorded predominantly in the Crocodile West and Marico catchment in terms of rivers and WWTWs. These sampling stations had tolerable to unacceptable levels of all or most of the selected physical and chemical water quality parameters and are of major concern for current and future aquatic health. Of great concern is that most WWTWs are of medium to high risk which indicates that these facilities greatly contribute to the overall degradation of the WMA. Rivers which are of medium to high risk are predominantly located downstream of these WWTWs as well as in some cases downstream of urban built-up or mining developments.

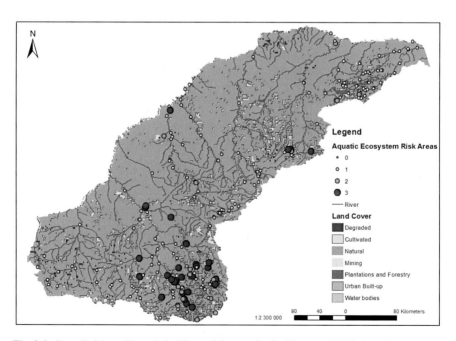

Fig. 9.4 Overall risk profile and significant risk areas for the Limpopo WMA (aquatic ecosystem standards)

9.1.4 Risk Areas for Irrigation

Irrigation water quality is predominantly of very low risk as 66% of sampling stations' measured water quality is of no risk and 26% are characterised as low risk. A total of 15 sampling stations were found to be of medium risk and 6 high risks (Fig. 9.5). Significant risk areas are once again located downstream of urban built-up areas or WWTWs. Sampling points which measured predominantly tolerable to unacceptable water quality and categorised as medium risk included 10 rivers located predominantly in the Crocodile West and Marico catchment. Five of the other medium risk sampling stations are located downstream of WWTWs, four of these are located in the Crocodile West and Marico catchment.

High-risk areas have been identified to be 5 river sampling stations and one WWTW located in the Crocodile West and Marico catchment. These river high-risk areas are once again predominantly located downstream of mining developments, WWTW or urban built-up areas. Farmers in the catchment, therefore, have to use caution when using water close to these points for irrigation purposes as it may have unintended negative effects on the growing of crops.

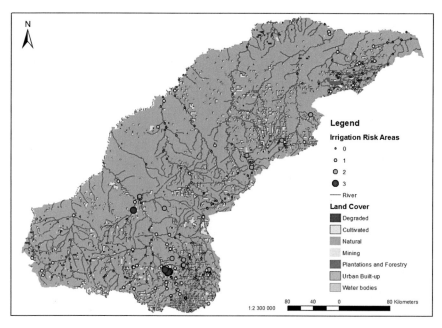

Fig. 9.5 Overall risk profile and significant risk areas for the Limpopo WMA (irrigation standards)

9.1.5 Risk Areas for Industrial Use

Water quality standards for industrial use is predominantly low to no risk as 42% of sampling stations were of low risk and 32% of no risk. However, a large amount of sampling stations measured tolerable to unacceptable industrial use water quality as a total of 49 sampling stations were found to be of medium risk (18%) and 17 sampling stations of high risk (8%) (Fig. 9.6). This may become a major concern in future with the continued increase in industrial and mining developments as tolerable to unacceptable water quality may negatively influence these sectors through affecting the efficiency of production processes and increasing financial costs.

Significant risk areas are spread across the WMA but are mostly located downstream of urban built-up areas, mining developments or WWTWs. Sampling points which measured predominantly tolerable to unacceptable water quality and categorised as medium risk located in the Crocodile West and Marico catchment included 21 rivers, 6 dams/barrages, 2 springs/eyes and 6 WWTWs. Medium-risk areas located in the Limpopo and Luvuvhu catchment areas comprised of 10 rivers, 1 spring/eye and 3 WWTWs.

High-risk areas have been identified to be 13 river-, 1 dam- and 3 WWTWs sampling stations located predominantly once again in the Crocodile West and Marico catchment. These high-risk areas are located within close proximity or directly downstream of mining developments, WWTWs or urban built-up areas. Industries, espe-

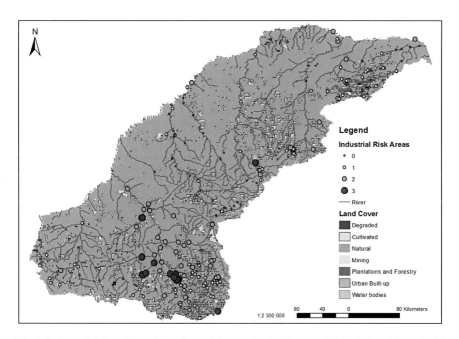

Fig. 9.6 Overall risk profile and significant risk areas for the Limpopo WMA (industrial standards)

cially in the Crocodile West and Marico catchment, therefore might have to use increased caution in future when directly using water from these points for industrial processes as it may have negative effects on production processes or cause unintended financial losses.

9.1.6 Chlorophyll a *and* Faecal Coliform *Risk Areas*

A total of only 36 sampling stations recorded *Chlorophyll a* concentrations during the time period. *Chlorophyll a* water quality standard for domestic use is predominantly of high to medium risk as 61% is of an unacceptable standard and 39% of sampling stations measured tolerable concentrations (Fig. 9.7). A large amount of sampling stations measured unacceptable to tolerable recreational water quality as a total of 22 sampling stations were found to have measured unacceptable levels (61%) and seven sampling stations measured tolerable levels (19%) (Fig. 9.8).

The Crocodile West and Marico catchment is once again the most degraded in terms of *Chlorophyll a* concentrations and high and medium risk areas are located close or directly downstream of urban built-up areas, agricultural areas as well as industrial activities.

Most sampling stations which measured *Chlorophyll a* during the time period, therefore, recorded unacceptable to tolerable concentrations. High concentrations

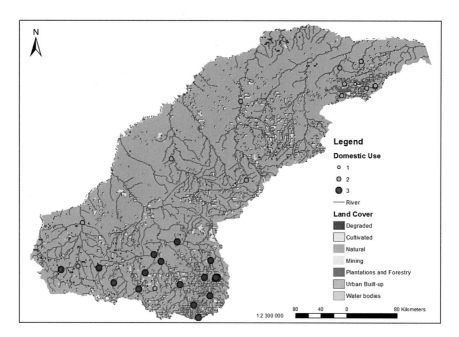

Fig. 9.7 Overall risk profile of *Chlorophyll a* for the Limpopo WMA (domestic use standards)

9.1 Limpopo WMA

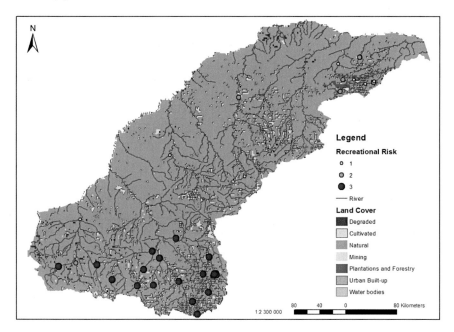

Fig. 9.8 Overall risk profile of *Chlorophyll a* for the Limpopo WMA (recreational use standards)

are due to tolerable to unacceptable levels of ammonia, nitrate as well as phosphate in these sampling areas mainly due to human activities. This should be of major concern as continued degradation will negatively influence all water use sectors through becoming unusable for domestic use, unsafe for recreational activities, further degrades aquatic ecosystems and negatively affects the efficiency of production processes and increasing financial costs for industrial activities ultimately affecting the water-food-energy nexus.

In terms of *Faecal coliform*, a total of 129 sampling stations measured its concentration within the WMA. Only three sampling stations measured acceptable standards and three sampling stations measured tolerable standards for domestic use. The other 123 sampling stations all measured unacceptable standards for domestic use and are spread across the WMA (Fig. 9.9).

In terms of irrigation use standards, 6 sampling stations measured acceptable standards, 40 tolerable standards and 83 unacceptable standards (Fig. 9.10). The predominant *Faecal coliform* risk profile for the WMA in terms of irrigation use is therefore high to medium and should be of concern for both domestic, recreational as well as irrigation use. The main contributor for the unacceptable standards of *Faecal coliform* can be attributed to poor or incompetent WWTW facilities as most of these facilities have been identified as problem areas in this WMA. Other contributors may include cultivated areas in the form of runoff of animal waste as well as from urban built-up or rural areas which may not have competent wastewater treatment facilities.

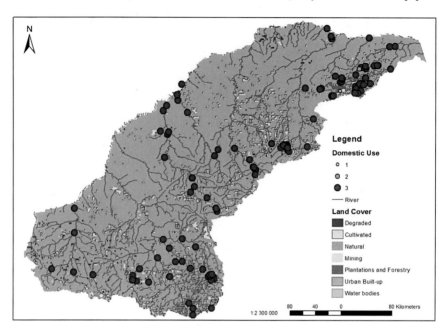

Fig. 9.9 Overall risk profile of *Faecal coliform* for the Limpopo WMA (domestic use standards)

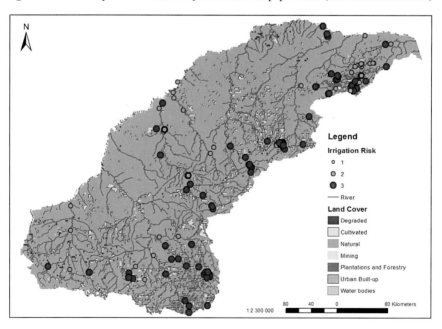

Fig. 9.10 Overall risk profile of *Faecal coliform* for the Limpopo WMA (irrigation standards)

Even though the WMA is largely underdeveloped the water resources of the WMA are heavily impacted in terms of salinity as well by urbanisation, wastewater discharges and platinum mining activities.

The Upper Marico River is in relatively good condition in terms of water quality. The agricultural return flow is a major factor and has caused the lower Marico falls to be of tolerable standard due to salinity issues. The areas which are monitored in the Limpopo and Luvuvhu catchments are of acceptable to ideal range of salinity but some areas are of unacceptable standard especially the upper Sand River catchment. The Crocodile West and Marico catchment should be of focus as most of the high and medium-risk areas are located here. The region needs to investigate the current state of their WWTWs as most of these do not comply with most or all water quality standards.

It should also be highlighted that a substantial portion of the water used in the catchment is transferred from the Vaal River and further afield and currently contributes to the good buffering capacity of the WMA. A decrease in additional water capacity could consequently lead to areas of acceptable water quality becoming tolerable to unacceptable due to the decrease in buffering capacity and affect all water uses in the WMA in terms of availability and quality.

9.2 Olifants WMA

9.2.1 WMA Overview

The Olifants WMA is made up of the Olifants, Letaba as well as Shingwedzi River catchments. The Shingwedzi River includes the Mphongolo, Phugwane, Shisha and Mashakwe Rivers and falls largely within the Kruger National Park. Land use outside the park area is predominantly subsistence farming and informal urban settlements. There are several small gold mines which have been developed in the south-western part of the catchment, however, these mines have very limited impact on the local economy and have recently been closed down.

The Letaba catchment, located in the northern region of the WMA, has two main tributaries namely the Klein and Groot Letaba Rivers (Fig. 9.11). The Groot Letaba River catchment includes main urban areas of Tzaneen and Nkowakowa whereas the Klein Letaba River catchment contains the town of Giyani. Rural populations are scattered throughout this catchment. The catchment is highly regulated by several dams in the upper and middle reaches of the river and is further regulated by irrigation weirs which limit flows into the Kruger National Park. The upper parts of the Klein Letaba River as well as along the Groot Letaba River are characterised by intense irrigation farming where vegetables and, citrus and a variety of fruits are grown. The already limited existing water resources have been overexploited to try and meet the growing demand for irrigation, afforestation, industries as well as domestic water demands.

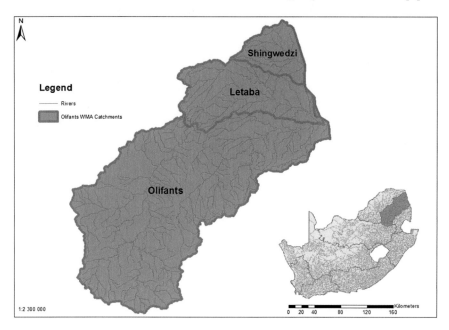

Fig. 9.11 Main catchments of the Olifants WMA

The Olifants catchment system is the major part of the WMA and its main tributaries include the Wilge, Elands and Ga-Selati Rivers on the west and the Klein-Olifants, Steelpoort, Blyde, Klaserie and Timbavati Rivers on the east. This catchment is highly utilised and regulated and its water resources are increasingly under pressure due to continued accelerated development and scarcity of water resources. Mining is the main economic activity within this catchment and extensive irrigation occurs. Most of the central and north-western areas of the catchment are characterised by underdevelopment and scattered rural villages with migrant workers. Rain-fed cultivation (grain and cotton) is dominant in the southern and north-western parts of the catchment. Most of the catchment, however, remains under natural vegetation for livestock and game farming as well as conservation. Severe overgrazing is, therefore, a major threat in many areas (Fig. 9.12).

The water quality monitoring of surface water resources is spread across the WMA but is highly concentrated in the south or Olifants catchment due to this catchment being highly developed in terms of mining, urban built-up and cultivation. A total of 278 sampling stations were evaluated (22 dams/barrages, 233 rivers and 23 WWTWs) distributed across the WMA.

9.2 Olifants WMA

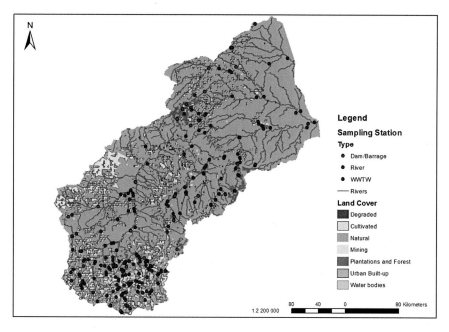

Fig. 9.12 Water quality sampling stations used and land cover of the Olifants WMA

9.2.2 Risk Areas for Domestic Use

The Olifants WMA is predominantly of no to low risk in terms of domestic use water quality standards especially in the central and north-eastern regions of the WMA. Of the 278 sampling stations, 180 (65%) of sampling stations fell in the no risk category and 68 (24%) were categorised as low risk (Fig. 9.13).

Significant risk areas (Risk level 2 and 3) are predominantly found in the Olifants catchment which is dominated by mining developments and urban built-up areas. High-risk areas are found in the Letaba catchment of the WMA, located downstream of WWTWs. A total of 28 medium risk areas (1%) were identified which are predominantly located in rivers and WWTWs located in the southern region of the WMA. Only 2 high-risk areas were identified which are both WWTWs type sampling stations. These 2 areas should, therefore, be highlighted as they are also located close to urban built-up areas, i.e. human populations which use the water for domestic purposes. Most of the risk areas are located within close proximity or directly downstream from mining developments (especially in the Olifants catchment), urban built-up or settlements as well as WWTWs.

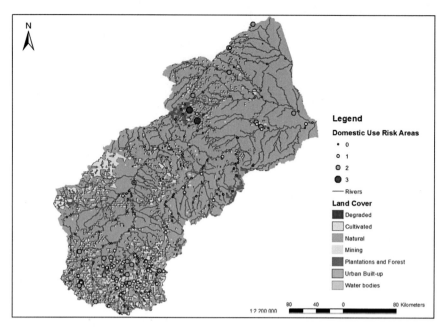

Fig. 9.13 Overall risk profile and significant risk areas for the Olifants WMA (domestic use standards)

9.2.3 Risk Areas for Aquatic Ecosystems

In terms of aquatic ecosystem water quality, the WMA is predominantly of low to medium risk in the less developed regions (Fig. 9.14). A total of 17 sampling stations were established to be of no risk and 199 (72%) of sampling stations were classified as low risk. Significant risk areas were once again predominantly located in developed regions especially in the Olifants catchment. A total of 41 medium risk areas and 21 high-risk areas were identified. These risk areas are mainly located close or downstream of mining developments or urban built-up. High-risk areas are mainly located downstream of WWTWs. Unacceptable concentrations of chloride, nitrate, ammonia and phosphate were recorded.

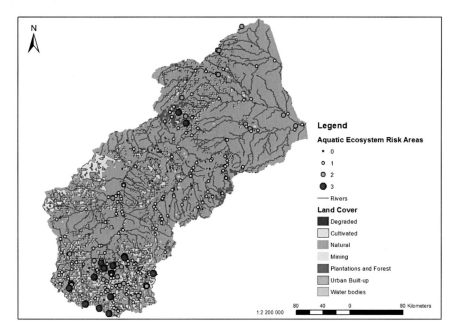

Fig. 9.14 Overall risk profile and significant risk areas for the Olifants WMA (aquatic ecosystem standards)

9.2.4 Risk Areas for Irrigation Use

Irrigation risk areas almost correspond with domestic use risk areas. The WMA is dominated by no- (189 sampling stations) and low (66 sampling stations) risk areas which mainly occur within the central and north-eastern regions of the WMA (Fig. 9.15). Low-risk areas are mostly located further downstream from cultivated land, mining developments as well as rural settlements.

Medium and high-risk areas were once again established downstream or in close proximity of mining developments, urban built-up as well as WWTWs. High-risk areas are WWTWs located in the north of the WMA close to scattered built-up areas. The Olifants catchment holds most of the risk for the WMA due to most of the medium risk areas falling in it. The high concentration of mining developments as well as urban built-up areas and WWTWs are the biggest contributors to pollution in this catchment. Other catchments also have medium to high-risk areas but to a much lesser extent. This is due to less development and main contributors are rural settlements and cultivation practices. Farmers which are located close to or downstream of mines, as well as urban built-up and WWTWs, therefore need to reserve caution as irrigation water quality risks have been identified.

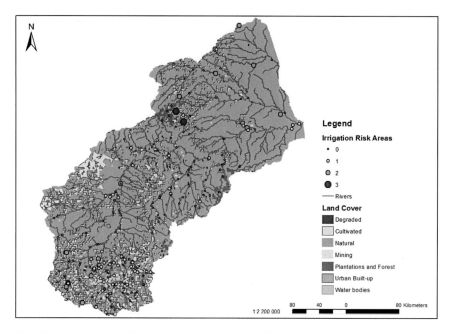

Fig. 9.15 Overall risk profile and significant risk areas for the Olifants WMA (irrigation standards)

9.2.5 Risk Areas for Industrial Use

The Olifants WMA is predominantly characterised by low-risk areas in terms of industrial use water quality (Fig. 9.16). A total of 67% sample stations were classified as low risk, followed by 16% no risk (located mainly in undeveloped areas), 14% medium risk and 3% high risk.

Medium risk areas are spread across the WMA. These areas are mostly located in close proximity of mining activities, urban built-up or rural settlements as well as downstream of WWTWs. The 10 sampling stations which were identified as high-risk areas are 2 WWTWs located in the Letaba catchment (north) and 8 river sampling points all located in the Olifants catchment. These sampling stations had tolerable to unacceptable levels of all or most of the selected physical and chemical water quality parameters and are of major concern.

Industrial activities which are located close to these areas and which make use of this water should take note as the use of poor water quality can negatively influence production processes and lead to unintended financial costs due to ineffective production processes.

9.2 Olifants WMA 199

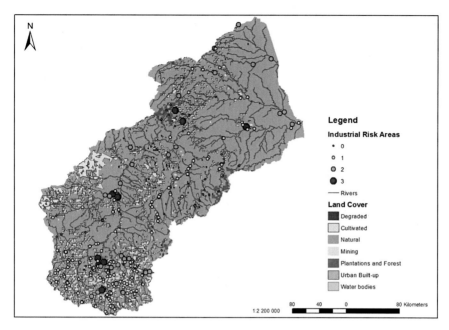

Fig. 9.16 Overall risk profile and significant risk areas for the Olifants WMA (industrial use standards)

9.2.6 Risk Areas for Chlorophyll a and Faecal Coliform

Only 31 sampling stations measured *Chlorophyll a* within the Olifants WMA. The *Chlorophyll a* concentrations in terms of domestic use are mainly of a medium risk as 84% of sampling stations were of tolerable standard (Fig. 9.17). Only 1 station measured acceptable domestic standards in the WMA and is in an underdeveloped area of the Olifants catchment.

In terms of recreational risk, the Olifants WMA is predominantly of a low risk as 61% of sampling stations measured acceptable standards for *Chlorophyll a* (Fig. 9.18). High-risk areas are however present. High-risk areas which measured unacceptable standards of *Chlorophyll a* can be found downstream or within close proximity of mining developments, urban built-up as well as WWTWs.

The risk profile of *Chlorophyll a* might change if more monitoring sampling stations are included especially in the Olifants catchment which is highly developed and degraded. A total of 126 sampling stations measured *Faecal coliform* levels within the WMA. The *Faecal coliform* levels for domestic use is overall of unacceptable standard as 97% of stations measured unacceptable levels. Only 2 stations measured acceptable levels and another 2 stations tolerable concentrations (Fig. 9.19).

In terms irrigation standards the WMA is of a high to medium risk as 71% of sampling stations measured unacceptable standards and 28% tolerable standards (Fig. 9.20). The Olifants WMA is therefore predominantly of high risk for water use

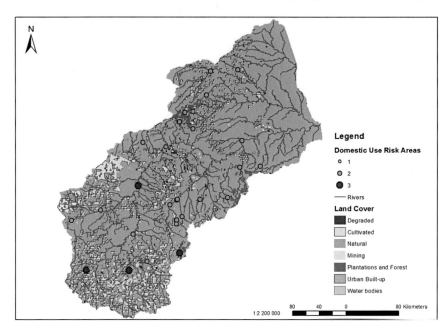

Fig. 9.17 Overall risk profile of *Chlorophyll a* for the Olifants WMA (domestic use standards)

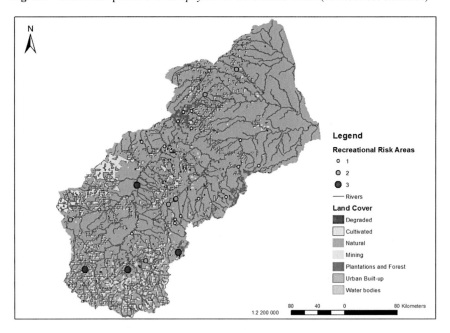

Fig. 9.18 Overall risk profile of *Chlorophyll a* for the Olifants WMA (recreational use standards)

9.2 Olifants WMA

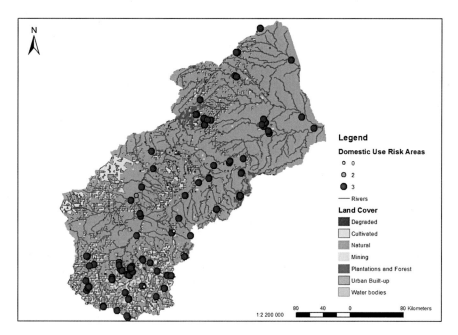

Fig. 9.19 Overall risk profile of *Faecal coliform* for the Olifants WMA (domestic use standards)

in terms of *Faecal coliform* levels. Tolerable to unacceptable levels of can mainly be attributed to runoff of animal wastes from cultivated areas as well as poor or inadequate WWTWs for urban built-up areas.

The Olifants WMA is, therefore, overall highly stressed and stress on its water resources will be exacerbated by continued population growth and development. The further development of its water resources is very limited and future developments will be forced to rely on local sources of water. Salinity is a major factor within this WMA as water sources especially within the Olifants catchment are mostly of unacceptable to tolerable standard. The Lower Olifants in the Kruger National Park is however of an acceptable standard as well as tributaries within the upper reaches of the catchment. The lower reaches of catchments are predominantly of unacceptable water standards and are largely due to mining, irrigation return flows as well as wastewater discharges. Smaller tributaries fall within the ideal range for most physical and chemical water quality parameters.

Water quality monitoring could be improved upon in the Letaba catchment (northern region) as it is currently limited. Monitoring of *Chlorophyll a* concentrations as well as *Faecal coliform* levels need to be improved upon as it is currently very limited and these parameters have been identified as being mostly of unacceptable to tolerable standards.

The ecological condition of the Olifants WMA falls in the moderately modified state to largely modified state. The upper reaches of the Olifants and within the Letaba and Shingwedzi catchments are predominantly natural to largely natural and

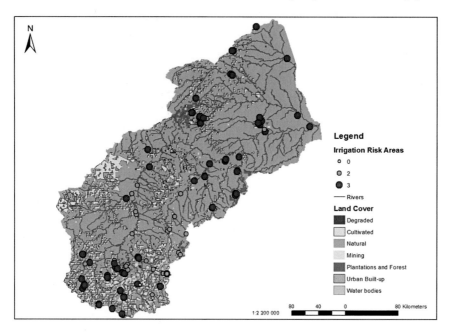

Fig. 9.20 Overall risk profile of *Faecal coliform* for the Olifants WMA (irrigation standards)

are less impacted as the majority of these rivers fall within the Kruger National Park. Mining and agricultural activities, as well as urban developments, are the biggest contributors to the modification of the WMA.

A precautionary approach to management is therefore required to try and maintain good conditions of some tributaries and attempt to minimise the further degradation of already severely stressed water resources. There are still surface water resources which have the capacity to accept degrees of impact, however, the development of these need to receive caution as their development can have unintended cumulative effects.

9.3 Inkomati-Usuthu WMA

9.3.1 Overview of WMA

The Inkomati-Usuthu WMA, located in the north-eastern part of the country, borders Mozambique and Swaziland and the Kruger National Park occupies 35% of the WMA. All of its rivers flows through Mozambique into the Indian Ocean and includes the Sabi-Sand River system, the Crocodile River East system, the Komati and Lomati system as well as the Usuthu River system (Fig. 9.21).

9.3 Inkomati-Usuthu WMA

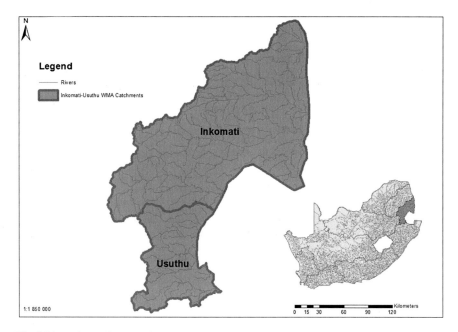

Fig. 9.21 Main catchments of the Inkomati-Usuthu WMA

Economic activities within the WMA are predominantly focussed on irrigation and afforestation with related industries and commerce. Due to the Kruger National Park, there is also a very strong ecotourism industry. Coal mining activities are, however, emerging in the upper reaches. The Kruger National Park remains the key feature of the WMA with the Sabi River flowing through it making it one of the most important ecological rivers in South Africa. Important urban centres in the WMA include Mbombela, White River, Komatipoort, Carolina, Badplaas, Barberton, Sabie, Bushbuckridge, Kanyamazan, Matsulu, Lothair, Piet Retief and Amsterdam. Dams have been constructed on all of the WMAs main rivers and tributaries, making the WMA well regulated. Water resources of rivers are consequently fully utilised or in balance and future water supply will require reconciliation options.

Joint management by South Africa and Swaziland exists in part of the water resources of the Komati Basin Water Authority. The Inkomati River is subject to international cooperative agreement with Mozambique which obligates South Africa to have a minimum of 2 m^3/s supplied to Mozambique. Swaziland is also dependent on the Usuthu River and relies on responsible upstream use from South Africa.

Groundwater utilisation is also relatively small due to the well-watered nature of the WMA. Most of the present yield from the Komati River, west of Swaziland, is transferred to the Olifants WMA for power generation.

Large areas of the WMA have been developed under irrigation and crops include fodder, grain, tobacco, citrus, tropical fruits and sugar. Commercial forestry is also present in the high rainfall escarpment and mountain areas of the WMA. Land outside of the Kruger National Park remains predominantly under natural vegetation for livestock and game farming as well as for conservation.

Overgrazing is rampant especially in some of the densely populated rural areas. Areas which have good soils and favourable topography make use of dryland cultivation. Mbombela, previously known as Nelspruit, is the biggest urban centre and scattered rural villages with high population densities are widespread in the WMA.

The upper parts of the Sabi River catchment are densely commercially afforested and the land use in the middle reaches is a combination of sub-tropical fruits and dense informal settlements. The lower reaches of the river falls within the Kruger National Park. The upper region of the Usuthu River catchment is sparsely populated and land use is dominated once again by afforestation with limited irrigation (Fig. 9.22).

Water quality monitoring of surface water resources is spread across most of the WMA except in the north-eastern part of the WMA where the Kruger National Park is located (Fig. 9.22). A total of 234 sampling stations had suitable data for the time period. The types of sampling stations include 21 dam/barrages, 185 rivers, 4 springs/eyes and 24 WWTWs spread across the whole WMA.

The current main stressors in the WMA are the high water demands by Eskom, irrigation, afforestation as well as industry and rapidly increasing domestic water demands. Mining is also a factor in the WMA. Major mining activities occur within the Inkomati catchment in the Baberton and Mbombela areas as well as in the Crocodile River catchment close to the Kaap River. Minerals include gold, asbestos, iron, nickel, copper and manganese and a significant number of coal reserves. Gold

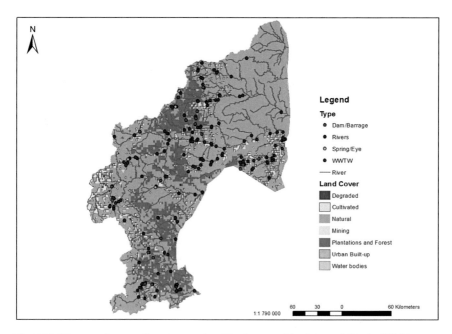

Fig. 9.22 Water quality sampling sites used and land cover of the Inkomati-Usuthu WMA

9.3 Inkomati-Usuthu WMA

and other mineral mining operations are widespread but have been reduced to small-scale operations. Coal mining occurs extensively in the south-west region of the WMA which is mainly used for fuel for large thermal power stations in the neighbouring Olifants WMA.

9.3.2 Risk Areas for Domestic Use

Inkomati-Usuthu WMA is of no to low risk in terms of domestic use water quality standards for physical and chemical water quality parameters. A total of 193 (82%) sampling stations were identified to be of no risk and 26 (11%) sampling stations of low risk (Fig. 9.23).

Fifteen sampling stations were recorded to be of medium risk. Three of these medium risk areas are rivers located within the Inkomati catchment primarily downstream of urban built-up areas as well as cultivated areas. Three of the WWTWs located in the Inkomati catchment have also been identified to be of medium risk. Seven river sampling points and 2 WWTWs within the Usuthu catchment were also classified as being of medium risk. The medium-risk rivers are once again located within close proximity or downstream of urban built-up as well as cultivated areas.

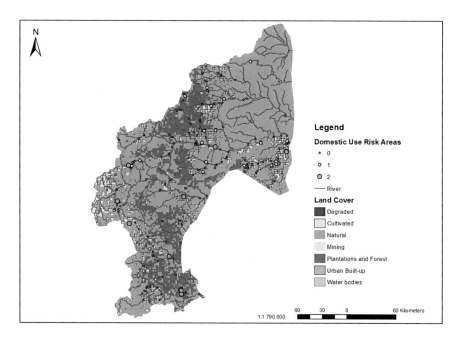

Fig. 9.23 Overall risk profile and significant risk areas for the Inkomati-Usuthu WMA (domestic use standards)

Mining operations located close to Baberton (western region of the catchment) also play a role as some medium-risk areas have also been identified.

Overall the WMA is of no to low risk in terms of domestic use water quality guidelines, however, this may change if precaution is not taken in future developments such as expansion of mining in the WMA.

9.3.3 Risk Areas for Aquatic Ecosystems

The WMA is mostly dominated by low-risk areas in terms of aquatic ecosystem water quality standards. A total of 167 (71%) sampling stations are classified as low-risk areas, followed by 36 (15%) of sample stations no risk, 20 (9%) medium risk and 11 (5%) high-risk areas (Fig. 9.24). Medium-risk areas are spread across the WMA, however, most occur within close proximity or directly downstream of urban built-up or cultivated areas. Five medium-risk areas are located at rivers and 6 downstream of WWTWs within the Inkomati catchment. The other medium-risk areas are 6 rivers and 3 of the 4 WWTWs located in the Usuthu catchment.

In terms of high-risk areas, 2 rivers in the Inkomati catchment and nine WWTWs fell in this risk category. These high-risk areas are located close to urban centres, rural settlements, cultivated areas as well as mining developments. Most of the sam-

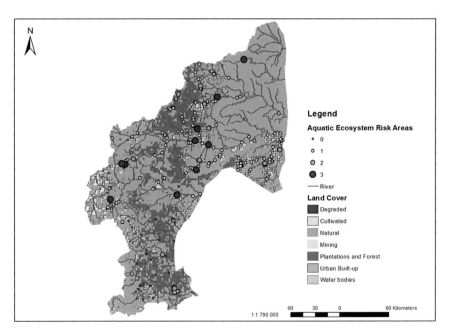

Fig. 9.24 Overall risk profile and significant risk areas for the Inkomati-Usuthu WMA (aquatic ecosystem standards)

9.3 Inkomati-Usuthu WMA

pling stations measured tolerable to unacceptable standards of EC, chloride, sodium, ammonia, phosphate and nitrate.

9.3.4 Risk Areas for Irrigation Use

The WMA is characterised by mostly acceptable irrigation water quality standards. Most sample stations were categorised as being of no to low-risk areas. A total of 198 (85%) are of no risk and 23 (10%) of low risk (Fig. 9.25). Only 13 sampling stations were classified as being of medium risk. These medium risk sample points are predominantly rivers and 2 WWTWs located in the Usuthu catchment. The tolerable concentrations can be attributed to cultivated areas as well as urban built-up. Tolerable concentrations were mostly recorded for ammonia, nitrate and phosphate water quality parameters which can be connected to fertiliser use and pollution from WWTWs. The overall water quality risk profile for irrigation use in the WMA is however no to low risk.

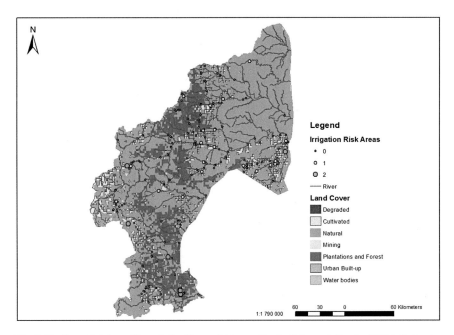

Fig. 9.25 Overall risk profile and significant risk areas for the Inkomati-Usuthu WMA (irrigation standards)

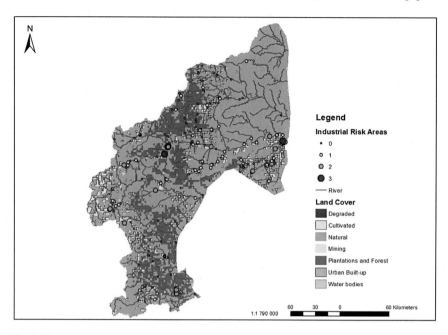

Fig. 9.26 Overall risk profile for the Inkomati-Usuthu WMA (industrial standards)

9.3.5 Risk Areas for Industrial Use

The overall water quality risk profile for the WMA in terms of industrial use ranges between no to low risk. The WMA therefore has mostly acceptable industrial water quality standards with 139 (59%) sample stations being of no risk and 70 (30%) sample stations categorised as low risk (Fig. 9.26).

Significant risk areas were established for the WMA. A total of 22 medium risk and 3 high-risk areas were identified in the WMA. Medium-risk areas are spread across the WMA. Most of the medium risk areas are rivers located in the Inkomati catchment close to urban built-up areas as well as cultivated areas. In terms of the 3 identified high-risk areas, all 3 are rivers located in the Inkomati catchment predominantly downstream of urban built-up areas.

9.3.6 Risk Areas for Chlorophyll a and Faecal Coliform

Concerningly, only 9 sampling stations measured and recorded *Chlorophyll a* concentrations within the WMA. There is therefore a major lack of *Chlorophyll a* monitoring within the WMA. All 9 of the sampling stations measured tolerable concentrations

9.3 Inkomati-Usuthu WMA

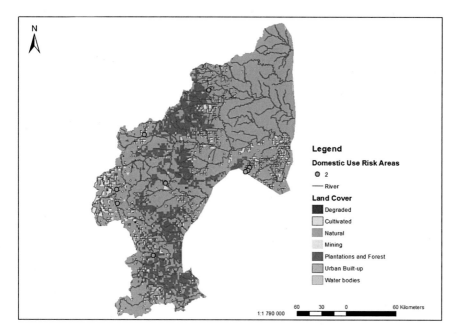

Fig. 9.27 Overall risk profile of *Chlorophyll a* for the Limpopo WMA (domestic use standards)

for domestic use and are located within close proximity or directly downstream of urban built-up or rural settlements (Fig. 9.27).

The WMA is overall of low risk in terms of recreational water quality standards as all sampling stations measured acceptable *Chlorophyll a* concentrations. No concrete conclusions can, however, be made due to the lack of available sampling stations in the WMA.

A total of 194 sampling stations measured *Faecal coliform* levels in the WMA for the time period and are scattered over the whole WMA. Monitoring is, however, lacking in the central region of the WMA. The WMA is overall of a high risk in terms of domestic use water quality standards as 181 (93%) sampling stations measured unacceptable *Faecal coliform* levels (Fig. 9.28).

The overall unacceptable levels for domestic use of *Faecal coliform* across the WMA can be attributed to animal wastes from cultivated areas but also poor or inadequate WWTWs. The scattered nature of rural settlements is also a challenge regarding the treatment of wastewater and contributes to further degradation in the form of sewage pollution.

The overall risk profile in terms of irrigation use is medium to high as 69% of sampling points measured tolerable levels of *Faecal coliform* (Fig. 9.29). High-risk areas are scattered across the catchment but are predominantly located either within close proximity or downstream of WWTWs, urban built-up, cultivated areas or rural settlements. The WMA needs to invest in the upgrading or development of WWTWs especially in terms of rural settlements.

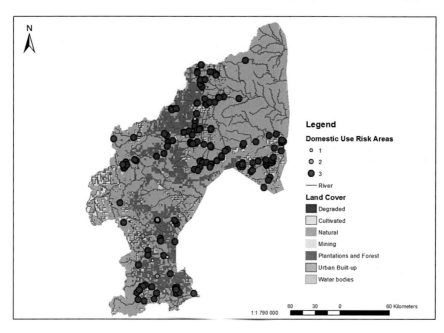

Fig. 9.28 Overall risk profile of *Faecal coliform* for the Limpopo WMA (domestic use standards)

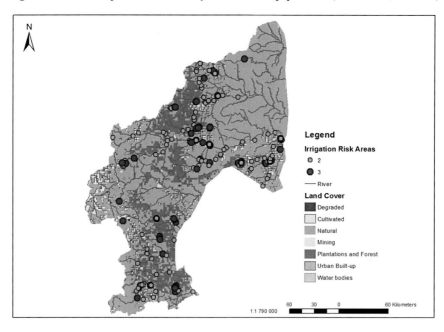

Fig. 9.29 Overall risk profile of *Faecal coliform* for the Limpopo WMA (irrigation use standards)

The overall ecological condition of the WMA is therefore mostly good to fair as most of the system is in a natural to largely natural state or moderately modified state. The lower reaches of the Crocodile River are however largely modified due to developments and is largely affected by acid rock drainage from old gold mining areas. Other smaller tributaries such as the Upper Sabie River and lower reaches of the Komati River out of Swaziland are also largely modified.

The water quality of the Inkomati-Usuthu catchment is therefore of a low risk in terms of physical and chemical water quality parameters. The salinity of the Crocodile river is tolerable with lower reaches falling in the unacceptable range. Some areas have, however, been identified as medium to high risk and should be taken note of especially those located downstream of WWTWs. The WMA clearly has an issue regarding *Faecal coliform* which should be addressed through upgrading or establishing WWTWs. The WMA should also invest in the expansion of sampling points for *Chlorophyll a* as currently monitoring is lacking.

In terms of water stress, 24% of the WMA is characterised by stressed surface water resources which are under threat and 3% of surface water resources which need a precautionary approach in management. Major current and future threats within the WMA therefore include future population growth, expansion of urban built-up areas and rural settlements as well as proposed mining developments.

9.4 Conclusions

The northern WMAs vary significantly according to the established risk areas. The Limpopo WMA is characterised by low rainfall and significant inter-dependencies for water resources between catchments and neighbouring WMAs. Significant risks were identified for most of the evaluated water quality standards in terms of the selected physical and chemical water quality parameters. The WMA is predominantly of low to no risk for most of the selected water quality parameters in terms of physical and chemical water quality parameters. However, significant risk areas were established and most of these were found to occur in the Crocodile West and Marico catchment which is also the second most populated catchment and largest proportional contribution to the country's national economy. These identified significant risk directly correlate with the highly altered nature of the catchment. Significant risk areas, especially those located in the Crocodile West and Marico catchment need to be addressed as this catchment is the second most populated in the country and plays a significant role in the country's economy. The further degradation of its water resources could pose increasing risks to different water use sectors in terms of quality but also availability as water may become unusable for certain uses.

The Olifants WMA is largely characterised by scattered rural populations, intense irrigation farming in the northern parts of the WMA and extensive mining, irrigation and urban areas in the south. Most of the catchment remains under natural vegetation for livestock and game farming as well as conservation. Severe overgrazing is a major threat in many areas and contributes to environmental degradation in the WMA together with extensive mining operations and other activities in urban built-up areas. The WMA is predominantly of low to no risk for most of the selected water quality parameters in terms of physical and chemical water quality parameters. The WMA is of medium risk for *Chlorophyll a* concentrations in terms of domestic use, however, the WMA has a low amount of sampling stations which recorded *Chlorophyll a*. The overall risk profile for *Chlorophyll a* might change (be of higher risk) if more monitoring stations are included. *Faecal coliform* levels were predominantly of an unacceptable standard and should be addressed by evaluating current WWTWs in terms of capacity, overall condition as well as management thereof.

The Inkomati-Usuthu WMA is predominantly focused upon irrigation and afforestation with related industries and commerce with coal mining emerging in the upper reaches. Current main stressors in the WMA are the high water demands by Eskom, irrigation, afforestation as well as industry and rapidly increasing domestic water demands. The WMA is predominantly of low to no risk for most of the selected water quality parameters in terms of physical and chemical water quality parameters. The WMA had only nine sampling stations which recorded *Chlorophyll a* and no concrete conclusion can be made due to lack of data. Mostly unacceptable levels of *Faecal coliform* were recorded for the whole WMA.

All of the northern WMAs need to expand their water quality monitoring network especially in terms of the measurement of *Chlorophyll a*, as eutrophication is an identified major water quality problem in the country, and *Faecal coliform*. All three of the WMAs also need to place a significant focus on the improvement of their WWTWs are these facilities are not functioning up to standard. These significant risk areas need to be addressed to minimise or limit future environmental degradation, significant human health risks as well as socio-economic costs.

Chapter 10
Current Water Quality Risk Areas for Vaal, Pongola-Mtamvuna and Orange WMAs

The focus is placed on the WMAs located in the central region of the country, namely the Vaal, Pongola-Mtamvuna and Orange WMAs. These WMAs are predominantly of low to no risk, however, numerous significant risk areas were established for all of the WMAs and directly correlate with the extent of modification of water sources or areas.

Significant risk areas were predominantly established downstream or within close proximity of urban centres, cultivated areas, mining developments as well as WWTWs. Anthropogenic activities play a major role in the degradation of water resources. WWTWs are especially of great concern for all of the WMAs as it is dominated by unacceptable to tolerable levels of most or all selected water quality parameters especially in terms of *Faecal coliform*. Most of the WWTWs facilities within these WMAs do not comply with set standards and can be attributed to these facilities being mismanaged, inadequate or in need of proper maintenance or upgrading. Proper sewage facilities are also needed to be developed for rural settlements. The WMAs could invest in the reuse of wastewater after it has invested in the upgrading or maintenance of current WWTWs as the reuse of wastewater could be seen as an untapped water source which could lessen pressure on catchments which are already experiencing a water deficit. Issues need to be addressed to avoid future significant environmental human health problems and risks especially within the Vaal and Pongola-Mtamvuna WMAs and the Upper Orange catchment due to the high concentration of the country's population and economic activities.

10.1 Vaal WMA

10.1.1 WMA Overview

The Vaal WMA includes the Upper, Middle and Lower Vaal catchments and its water resources are of great significance for the country and its population as it supports major economic activities and a population of approximately 12 million people. The Vaal River system stretches from Ermelo in the north-east to Vryburg in the North West to Douglas in the south-west to Harrismith in the east (Fig. 10.1).

The Vaal River is the primary water resource in the Vaal system with numerous significant tributaries. The Vaal River flows 1,415 km originating at the Sterkfontein beacon flowing southwest to the Orange River confluence. It is deemed to be the country's and southern African region's most developed and regulated river with 90 major man-made impoundments located on the main stem and tributaries. The Vaal River system is characterised by extensive water resource infrastructure and is linked to substantial water transfer systems to other water resource systems such as Thukela, Usuthu and Lesotho. Significant transfers are also made out of the Upper Vaal catchment through the Rand Water distribution system to the Crocodile West and Marico catchments. The system's supply reaches most of the Gauteng Province, Eskom's power stations as well as Sasol's plants, the North West and Free State goldfields, the North West platinum and chrome mines, the Northern Cape iron and

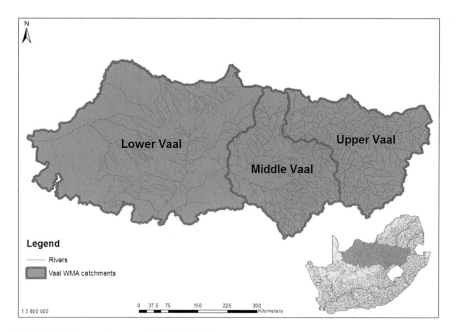

Fig. 10.1 Main catchments of the Vaal WMA

10.1 Vaal WMA

manganese mines, Kimberley, and several other small towns along the course of the river as well as numerous large irrigation schemes.

The Upper Vaal catchment's water resources within the catchment is therefore largely developed and highly altered by developments. The Middle Vaal has few major development centres with agriculture and mining being the primary activities. The Lower Vaal is also less developed with agriculture being the predominant land use. Significant types of development in the system include both formal and informal urbanisation, industrial growth, agricultural activities and widespread mining activities. The development of the system has led to the widespread deterioration of water quality which has required management interventions to ensure that water is of an acceptable quality for all users in the system especially as activities have continued to grow and intensify. The salinization and eutrophication are the two major water quality problems which the WMA has been experiencing.

The Upper Vaal catchment is characterised by extensive urban, mining and industrial areas in the northern and western parts. The urbanised area is primarily located in the Gauteng Province and extends beyond the WMAs boundary. Other developments in the catchment are related to dryland agriculture. The catchment also includes numerous large towns situated around mining, industrial and agricultural development areas and mining plays a significant role in the area's economy.

The Middle Vaal catchment has been shaped by the discovery of diamonds in the North West area which is now dominated by gold mining. The central parts of the catchment are dominated by extensive dryland cultivation. The primary urban areas include Welkom, Klerksdorp and Kroonstad. Irrigation is practiced mainly downstream of dams located along the main tributaries and locations along the Vaal River.

The Lower Vaal catchment is characterised by widespread livestock farming as the main activity and large-scale dryland cultivation in the north and east due to the arid climate. Intensive irrigation occurs at Vaalharts as well as other locations along the Vaal River and the most significant urban area is Kimberley in the south of the catchment. The catchment is also characterised by several towns and scattered rural settlements which are primarily found in the central and eastern parts.

The water quality monitoring of surface water resources is very limited within the Middle and Lower Vaal (Fig. 10.2). This might be attributed to these catchments being less developed and primary focus being placed on the Upper Vaal catchment due to its significance to various water sectors and the country. A total of 382 sampling stations qualified and were used for the analysis. Of these 382 sampling stations, 39 are dams/barrages, 221 rivers, 7 springs/eyes, 1 wetland and 114 WWTWs sampling points. All of these sampling points primarily occur within the Upper Vaal WMA.

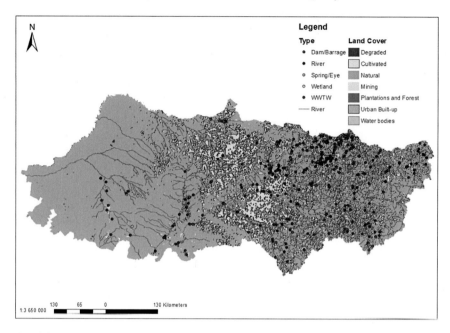

Fig. 10.2 Water quality sampling sites used and land cover of the Vaal WMA

10.1.2 Risk Areas for Domestic Use

The Vaal WMA is predominantly of no to low risk in terms of domestic use water quality standards. Most sampling stations measured acceptable water quality standards in terms of the selected physical and chemical water quality parameters (Fig. 10.3). Of the 382 sampling stations, 206 (54%) sampling stations are categorised as no risk, 96 (25%) as low risk, 39 (13%) as medium risk and 29 (8%) as high risk.

Significant risk areas (risk level 2 and 3), are in close proximity or directly downstream from urban built-up, mining, as well as cultivated land cover and, are spread across the WMA. Medium-risk areas include 4 rivers and 18 WWTWs in the Middle Vaal catchment, 17 WWTWs in the Upper Vaal catchment and 2 rivers and 10 WWTWs in the Lower Vaal catchment. Medium-risk areas are therefore predominantly within close proximity or directly downstream of WWTWs.

High-risk areas constitute of 2 rivers and 13 WWTWs in the Middle Vaal catchment, 1 river and WWTWs in the Upper Vaal catchment and 12 WWTWs in the Lower Vaal catchment.

WWTWs are therefore not up to standard in terms of domestic use water quality as most of these are either of medium to high risk. It should be noted that none of the WWTWs in the Middle Vaal catchment measured acceptable levels and are all either of medium to high risk. This should be of great concern to the whole WMA as these wastewaters can have widespread effects for all water users.

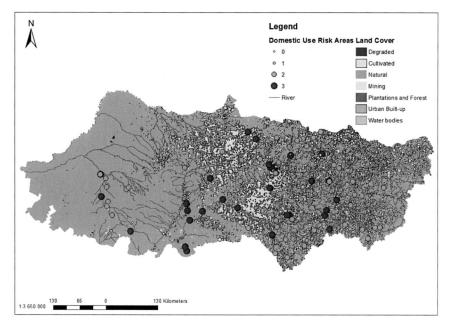

Fig. 10.3 Overall risk profile and significant risk areas for the Vaal WMA (domestic use standards)

10.1.3 Risk Areas for Aquatic Ecosystems

The current aquatic ecosystem water quality is predominantly of low to high risk. Most sampling points measured acceptable standards, however, a large amount also recorded tolerable to unacceptable standards. A total of 242 (63%) sampling stations were classified as low-risk areas, 65 (17%) as medium risk, 61 (16%) as high risk and the remaining 14 sampling stations as no risk (Fig. 10.4).

Significant risk areas are once again spread across the WMA. Medium-risk areas include 3 rivers and 18 WWTWs in the Middle Vaal catchment, 1 dam, 11 rivers and 20 WWTWs in the Upper Vaal catchment and 2 rivers and 10 WWTWs in the Lower Vaal catchment.

High-risk areas include 9 rivers and 13 WWTWs in the Middle Vaal catchment, 2 rivers and 25 WWTWs in the Upper and 12 WWTWs in the Lower. WWTWs once again pose a significant risk, especially in the Middle Vaal catchment where all WWTWs were found to have tolerable to unacceptable water quality standards in terms of the selected physical and chemical water quality parameters. Other risk areas occur downstream or within close proximity of urban built-up as well as mining developments. Cultivated areas were found to also be a contributing factor. Most of these sampling stations measured unacceptable to tolerable levels of ammonia, nitrate and phosphate levels.

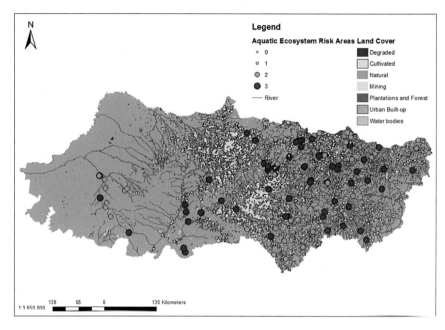

Fig. 10.4 Overall risk profile and significant risk areas for the Vaal WMA (aquatic ecosystem standards)

10.1.4 Risk Areas for Irrigation Use

The irrigation water quality of the Vaal WMA is mostly of no to low risk (Fig. 10.5). Of the 382 sampling stations, 213 (56%) sampling stations were classified as no-risk areas as most of the measured water quality parameters were of acceptable standard. A total of 92 (24%) sample stations are of low-risk, 49 (13%) medium-risk areas and 28 (7%) high-risk areas.

In terms of medium-risk areas, these areas are spread across all three catchments. Medium-risk areas were established to be 5 rivers and 16 WWTWs in the Middle Vaal catchment, 18 WWTWs in the Upper- and 4 rivers, 1 dam and 5 WWTWs in the Lower. High-risk areas include 2 rivers and 13 WWTWs in the Middle Vaal catchment, 1 river and WWTW in the Upper- and 11 WWTWs in the Lower. Therefore, once again, WWTWs have been identified as the primary cause of risk within the Vaal WMA. Most of these sampling stations once again primarily recorded unacceptable to tolerable concentrations of EC, sodium, ammonia, nitrate and in some cases tolerable to unacceptable pH levels over the time period. Cultivated areas which use water for irrigation purposes, located within close proximity or downstream of these facilities should, therefore, use caution especially in the Middle Vaal catchment where all WWTWs have been classified as medium to high risk.

10.1 Vaal WMA

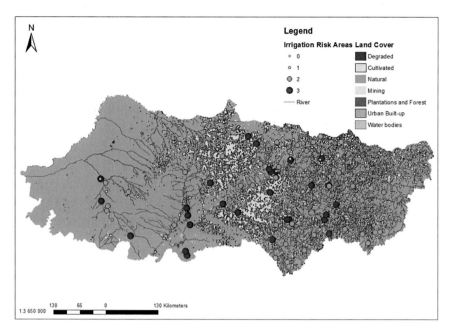

Fig. 10.5 Overall risk profile and significant risk areas for the Vaal WMA (irrigation standards)

10.1.5 Risk Areas for Industrial Use

The Vaal WMA is predominantly of low to medium risk in terms of industrial water quality. A total of 166 (43%) sampling stations have been classified as low risk followed by 109 (29%) medium risk, 67 (18%) high risk and 40 (10%) posing no risk (Fig. 10.6).

Medium-risk areas are predominantly located in the Upper Vaal catchment where 30 river sampling points, 4 dams, 1 spring/eye and 25 WWTWs were classified as being of medium risk for industrial water use. In the Middle Vaal catchment, 3 dams, 15 rivers and 12 WWTWs were classified as medium risk followed by the Lower Vaal catchment where 1 river, 1 dam and 17 WWTWs are of medium risk. Multiple high-risk areas were also identified in the WMA. These include 13 rivers and 7 WWTWs in the Middle Vaal catchment, 21 rivers, 3 dams and 1 WWTWs in the Upper and lastly 15 rivers, 3 dams and 4 WWTWs in the Lower. High-risk areas are therefore predominantly located in the Upper and Middle Vaal catchments, close to or downstream of WWTWs as well as urban built-up or mining developments. High and unacceptable concentrations of pH, EC, chloride and sulphate were measured over the time period.

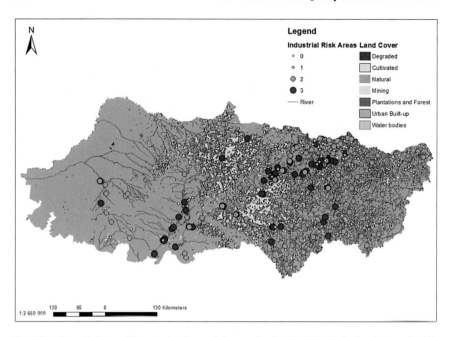

Fig. 10.6 Overall risk profile and significant risk areas for the Vaal WMA (industrial standards)

10.1.6 Risk Areas for Chlorophyll a and Faecal Coliform

The Vaal WMA only had 37 sampling stations which measured *Chlorophyll a* during the time period. This should be of concern as eutrophication has been identified as a significant water quality problem in the WMA which should receive increased monitoring in order to establish the extent of it accurately. *Chlorophyll a* water quality standards for domestic use is predominantly of a high risk as 30 of the 37 sampling stations were classified as high risk due to recording unacceptable concentrations of *Chlorophyll a*. The other seven sampling stations recorded tolerable concentrations and were classified as medium risk (Fig. 10.7). Most of the high-risk areas are located downstream or urban built-up areas as well as WWTWs, mining developments and cultivated areas.

In terms of recreational water quality, the Vaal WMA is once again predominantly of high risk (30 sampling stations) followed by some low (3 sampling stations) and medium (4 sampling stations) risk areas following the same trend as in the case of domestic use risk areas (Fig. 10.8).

For the Vaal WMA to improve upon or address its eutrophication water issues, the WMA will have to invest in the expansion of *Chlorophyll a* monitoring points to establish additional risk areas as well as probable causes for the problem. Currently, most of the sample stations have measured unacceptable levels of *Chlorophyll a* in terms of both domestic use and recreational standards which once again emphasises

10.1 Vaal WMA

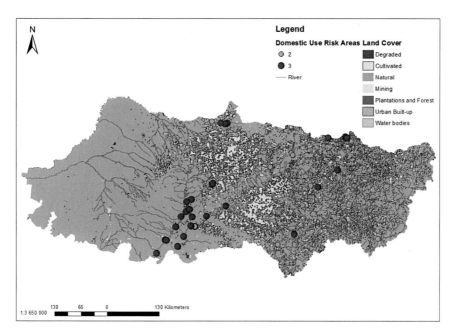

Fig. 10.7 Overall risk profile of *Chlorophyll a* for the Vaal WMA (domestic use standards)

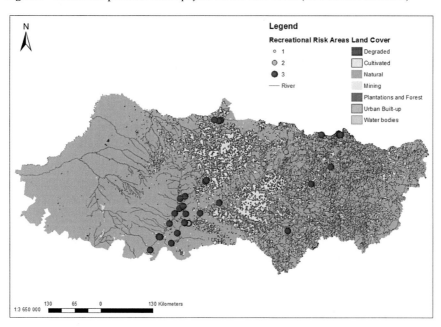

Fig. 10.8 Overall risk profile of *Chlorophyll a* for the Vaal WMA (recreational use standards)

how extensive eutrophication is within the WMA and should receive proper attention to reduce the problem.

The Vaal WMA also has an extensive *Faecal coliform* problem in terms of domestic use and irrigation water quality standards. A total of 339 sampling points recorded levels of *Faecal coliform* for the Vaal WMA. Of these 339, a total of 328 (98%) sampling stations recorded unacceptable levels of *Faecal coliform* for domestic use (Fig. 10.9).

In terms of irrigation water quality standards the WMA is not faring much better as 137 (40%) sample stations measured tolerable and 202 (60%) unacceptable levels of *Faecal coliform* (Fig. 10.10), which can be mainly attributed to poor or inadequate WWTWs.

The overall water quality of the Vaal WMA is concerning in terms of aquatic ecosystems, industrial as well as *Chlorophyll a* and *Faecal coliform* water quality standards. Salinity levels within the WMA are predominantly tolerable to acceptable which also emphasises that its water resources are under stress. The headwaters of the Vaal River are mostly of an acceptable standard. Some tributaries of the Vaal especially those which are located within close proximity or downstream of urban built-up, mining operations, as well as WWTWs, have poor water quality. The Blesbok Spruit tributary is a prime example. The water quality downstream of the Vaal Dam towards the Middle and Lower Vaal catchments are of a tolerable standard and are mainly impacted by flows from tributary catchments. Catchments which are

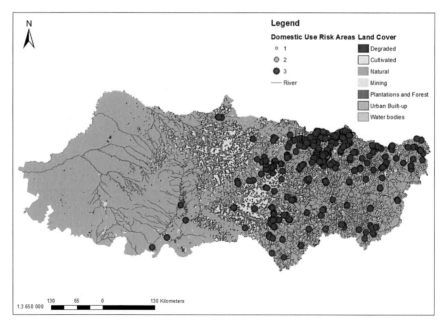

Fig. 10.9 Overall risk profile of *Faecal coliform* for the Vaal WMA (domestic use standards)

10.1 Vaal WMA

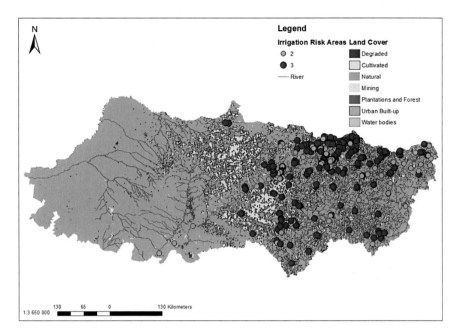

Fig. 10.10 Overall risk profile of *Faecal coliform* for the Vaal WMA (irrigation standards)

characterised by poor water quality and unacceptable ranges of salinity include the Suikerbosrand, Klip River and the Mooi River.

The Middle and Lower Vaal catchment are characterised by mostly tolerable standards. The Koekemoer Spruit and Schoon Spruit located in the Middle Vaal catchment are however of an unacceptable state especially in terms of salinity. Unacceptable levels of salinity are mainly attributed to mining operations in the catchment. The Lower Vaal deteriorates in terms of salinity standards downstream of the Bloemhof Dam. The Harts River's water quality is extremely poor especially in terms of salinity and contributes significantly to salinity problems in the lower Vaal River. Its poor state is largely due to irrigation return flows.

The overall ecological condition of the Vaal WMA ranges from a moderately modified condition to seriously modified state. This is largely due to most of its water resources being developed to capacity. The Upper Vaal catchment is described to be in a moderately modified condition except for the water resources within the Vaal Barrage catchment which are in poor condition and is largely so seriously modified. Other smaller tributaries in the headwater catchments of the Upper Klip and Upper Wilge Rivers are in a good ecological condition and in a largely natural present ecological state due to these areas not being affected by human developments or activities.

The Middle Vaal catchment's water resources range between moderately modified state to largely modified state with the small tributaries located in less developed areas being in a largely natural state. The lower reaches of the catchment's main tributaries are seriously modified and is indicative of unsustainable systems leading to large losses of biota and ecosystem habitats. The Lower Vaal catchment is classified as being in a largely modified ecological state downwards from the Bloemhof Dam with many of its tributaries being in a moderately modified state.

The WMA has also undergone the water balance reconciliation of the system, and has required the implementation of five core interventions to ensure sufficient water availability to users in the short term. Interventions have included the eradication of extensive unlawful water use, implementation of water conservation and water demand management measures, reuse of water, implementation of an integrated water quality management strategy and lastly the implementation of Phase two of the Lesotho Highlands Project. The surface water resources of the Vaal WMA can, therefore, be described to be stressed and under threat. Precautionary approaches in management should be followed to maintain some good condition.

10.2 Pongola-Mtamvuna WMA

10.2.1 WMA Overview

The Pongola-Mtamvuna WMA contains the Mhlatuze, Pongola, Mkuze, Mfolozi, Thukela, Mngeni, Mvoti, Mkomazi, Mtamvuna and Mzimkulu systems which vary in size from medium-to-large catchments with all rivers flowing towards the ocean apart from the Pongola River which joins the Maputo River in Mozambique. The WMA is characterised by some water transfer across catchments with the most important one being water transfer from the Thukela system to the Vaal system, reserving additional water for long-term requirements. The main catchments of the WMA are the Mhlatuze, Thukela and Mvoti catchments (Fig. 10.11).

The primary challenge facing the WMA is the additional water supply needed to meet growing needs of the Kwazulu-Natal Coastal Metropolitan Area which includes Durban-Pietermaritzburg, KwaDukuza in the North to Amanzimtoti in the South. Water requirements are constantly increasing and catchments are already in deficit. The Thukela pipeline project which entails the raising of Hazelmere Dam and the construction of Spring Grove Dam has been implemented as an intervention to address water shortages. The construction of dams on the Mkomazi and Mvoti Rivers, as well as desalinisation and reuse of wastewater and seawater desalinisation, are other options which are also being considered.

The Mhlathuze, Mfolozi, Mkuze and the Pongola catchment areas are characterised by industrial, agricultural and transportation as their main economic sectors. The land use in these catchment areas, in terms of water use, is predominantly irrigation and afforestation. Large proportions of the area is tribal land which is usually used for stock farming with old mining areas are located close to Vryheid.

The Richards Bay area can be described as a fast-growing industrial hub containing numerous industrial complexes within the Mhlatuze catchment. Majority of the

10.2 Pongola-Mtamvuna WMA

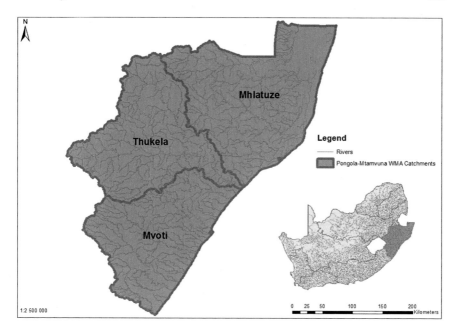

Fig. 10.11 Main catchments of the Pongola-Mtamvuna WMA

population in this catchment still live in rural areas. The Pongola System includes the massive Pongolapoort Dam upstream which serves sugarcane farmers with irrigation water. The Mkuze and Mfolozi sub-catchments are largely unregulated and support mainly forestry and irrigation water users. The world famous heritage site, Lake St Lucia, is located in this catchment. Upstream water use, poor catchment management and widespread erosion have negatively impacted the ecological condition of the St Lucia estuary, and there is still the potential for water resource development in the wider catchment area.

The largest river in the WMA is the Thukela River which includes the Little Thukela, Klip, Bloukrans, Bushmans, Sundays, Mooi and Buffalo rivers as major tributaries and forms the Thukela catchment. The river's water resources are primarily used to support water requirements in other parts of the country in the form of large water transfers to neighbouring catchments. These transfers include water transfers into the Vaal System, to the Mhlatuze catchment in the north and the Mooi-Mgeni system in the South. The catchment also holds eight major dams which include Woodstock, Spioenkop, Zaaihoek, Driel Barrage, Kilburn, Ntshingwayo, Craigie Burn and Wagendrift Dams. Major urban areas include the towns of Newcastle, Dundee, Ladysmith and Escourt. Most of the catchment's population is dependent on agriculture for their livelihoods and subsistence farming is practised on the communal land covering most of the catchment. In terms of industries, the catchment also includes the paper mill at Mandini.

The Mngeni, Mvoti, Mdloti, Mzimkulu and Mtamvuna systems form the southern portion of the WMA and the Mvoti catchment. The Mzimkulu and Mkomazi comprise the two larger river systems, the Mngeni and Mvoti the two medium-sized and the Mzumbe, Mdloti, Tongaat, Ifafa, Lovu and Mtamvuna as several smaller river systems. The Mvoti, Mdloti and Mngeni catchment areas are highly stressed due to water requirements exceeding available water supply. The catchment is the fourth largest contributor to the country's GDP and predominant land uses include major urban settlements along the Durban-Pietermaritzburg axis with the Durban metropolitan area being one of South Africa's major urban areas. Several small urban settlements in the locality support the surrounding agricultural sector. Large zones of commercial and subsistence agricultural land are located on the outskirts of urban areas. Dominant commercial agriculture includes timber, sugar cane, pastures and cash crops and there is substantial industrial development in main urban areas of Durban, Stanger and Pietermaritzburg. Additionally, it is important to note that the catchment is not facing any significant mining concerns or power stations.

Water quality monitoring of surface water resources are limited in some areas of the WMA (Fig. 10.12). Water quality monitoring is concentrated close to urban settlements and on the coastal area. A total of 375 sampling stations were evaluated. Of these 375 sampling stations, 39 are dams/barrages sampling stations, 208 river-, 89 estuary-, 2 wetland- and 37 WWTWs. The St Lucia estuary is intensely monitored as it is a world heritage site.

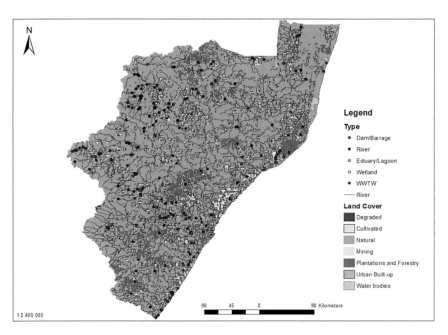

Fig. 10.12 Water quality sampling sites used and land cover of the Pongola-Mtamvuna WMA

10.2.2 Risk Areas for Domestic Use

Pongola-Mtamvuna WMA is mostly of no to low risk in terms of domestic use water quality. Most sampling stations measured acceptable levels for the selected physical and chemical water quality parameters. A total of 217 (58%) sampling stations were classified as no-risk areas, 107 (29%) low risk, 36 (10%) medium risk and 15 (3%) high risk (Fig. 10.13).

Significant risk areas are mostly located in the North West region of the WMA as well as close to coastal areas. Medium-risk areas include 1 river, 6 estuary and 3 WWTWs sampling areas in the Mvoti catchment, 4 rivers and 15 WWTWs in the Thukela catchment and lastly, 1 river, 1 lake and 5 WWTWs in the Mhlatuze catchment.

High-risk areas within the WMA include three estuary- and four WWTWs sampling stations in the Mvoti catchment, three river and two WWTWs in the Thukela- as well as two river and one estuary sampling points in the Mhlatuze catchment. Most of these risk areas are located downstream of urban built-up areas as well as cultivated land use. Tolerable to unacceptable levels of EC, calcium, chloride, sodium, ammonia, nitrate and phosphate were measured by these sampling stations. WWTWs once again play a key role in the degradation of the WMAs water resources as most of these are either of medium to high risk. Water resources which are located

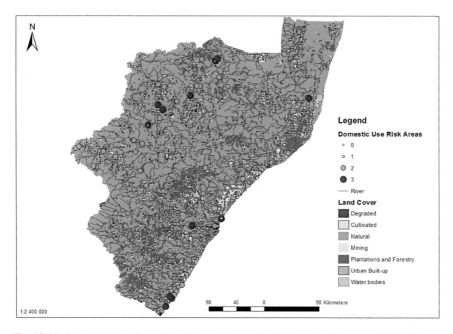

Fig. 10.13 Overall risk profile and significant risk areas for the Pongola-Mtamvuna WMA (domestic use standards)

downstream of these facilities are therefore a risk for domestic use and populations located close to these areas should refrain from using the water.

10.2.3 Risk Areas for Aquatic Ecosystems

The WMA is mostly of low risk in terms of aquatic ecosystem water quality standards (Fig. 10.14). A total of 259 (69%) sampling stations recorded mostly acceptable standards for the selected physical and chemical water quality parameters. No risk sampling stations include 72 (19%), medium-risk 33 (9%) and high-risk 11 (3%) sampling stations. The St Lucia estuary is predominantly of low risk in terms of aquatic ecosystem water quality standards. Risk areas are mostly the same as domestic use water quality.

Significant risk areas include mostly medium-risk areas which can be grouped as the following. Two estuary- and 3 WWTWs sampling stations in the Mvoti catchment, 6 river and 14 WWTWs in the Thukela catchment, 2 rivers and 5 WWTWs in the Mhlatuze catchment. In terms of high-risk areas, 3 estuary- and 4 WWTWs sampling stations in the Mvoti catchment, 3 WWTWs in the Thukela catchment and only 1 river sampling station in the Mhlatuze catchment recorded unacceptable standards. It

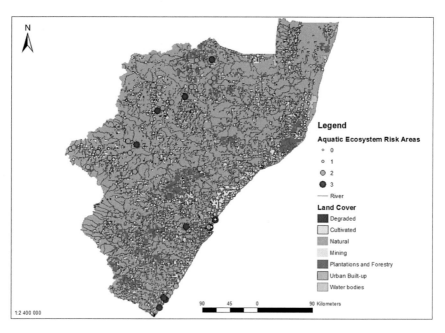

Fig. 10.14 Overall risk profile and significant risk areas for the Pongola-Mtamvuna WMA (aquatic ecosystem standards)

10.2 Pongola-Mtamvuna WMA

should be highlighted that in terms of the Mhlatuze catchment, all except 1 WWTWs is of medium to high risk.

Runoff from urban areas, especially from the Durban metropolitan area as well as wastewater from WWTWs across the WMA is of high risk. Cultivated areas also play a role in the degradation of the WMAs water resources however to a lesser extent. WWTWs, therefore, need to be investigated to ensure that wastewater is treated and up to standard to avoid further degradation of aquatic ecosystems. Even though the WMA is overall of low risk for aquatic ecosystems, this may become a larger issue if WWTWs are not addressed.

10.2.4 Risk Areas for Irrigation Use

Irrigation water quality is mostly of acceptable standard in the WMA. The WMA has a no-to-low-risk profile as of the 375 sampling stations, 218 (58%) sampling stations were classified as no risk and 121 (32%) as low risk (Fig. 10.15).

Sampling stations which measured mostly tolerable concentrations include 1 river, 5 estuary- and 7 WWTWs sampling stations in the Mvoti catchment, 2 river- and 14 WWTWs sampling stations in the Thukela catchment and 1 river-, 1 dam and 5 WWTWs in the Mhlatuze catchment. High-risk areas which mostly measured

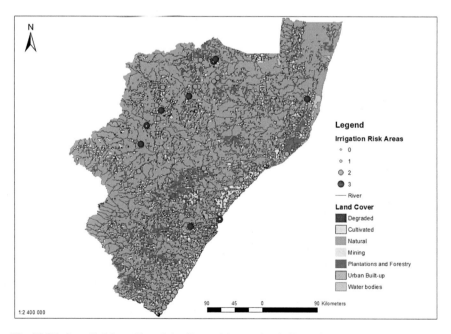

Fig. 10.15 Overall risk profile and significant risk areas for the Pongola-Mtamvuna WMA (irrigation standards)

unacceptable levels especially in terms of EC, calcium, chloride, sodium, ammonia, nitrate and phosphate include 1 estuary- and 1 WWTWs sampling station in the Mvoti catchment, 2 river- and 3 WWTWs sampling stations in the Thukela catchment and 2 rivers and 1 estuary sampling station in the Mhlatuze catchment.

High-risk areas are predominantly located downstream of WWTWs or urban built-up land cover as well as some cultivated areas. Cultivated areas located downstream of WWTWs or urban built-up areas should reserve some caution when using water for irrigation especially those located at the mentioned high-risk areas.

10.2.5 Risk Areas for Industrial Use

Industrial risk areas are predominantly of a low to no risk in the WMA. A total of 190 (51%) sampling stations were classified as low risk and 130 (35%) as having no risk. Medium-risk areas are spread across the WMA. A total of 45 sampling stations were classified as medium risk and measured mostly tolerable concentrations of the selected physical and chemical water quality parameters (Fig. 10.16).

The Mvoti catchment contains 3 river-, 7 estuary- and 4 WWTWs sampling points of medium risk and only 2 river sampling stations which were classified as high risk. In terms of the Thukela catchment, 8 rivers-, and 9 WWTWs sampling areas were

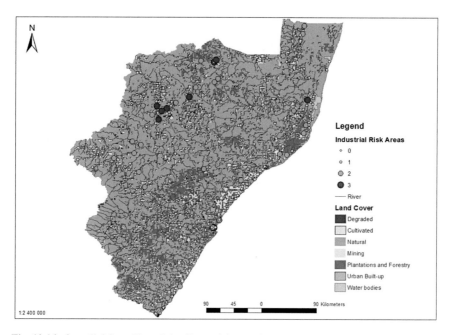

Fig. 10.16 Overall risk profile and significant risk areas for the Pongola-Mtamvuna WMA (industrial standards)

classified as medium risk and 3 river- and 2 WWTWs sampling points are high-risk areas. Lastly, 6 river-, 3 dam/lakes- and 5 WWTWs sampling points are medium-risk areas and only 2 river- and one estuary sampling point was classified as high risk.

The WWTWs within the Mhlatuze catchment are once again of concern as 5 of the 6 sampling points are identified as medium risk. Industrial activities located downstream of these facilities need to serve caution if water resources within these identified risk locations are used within their processes. Water resources downstream of urban built-up land cover may also pose a risk.

10.2.6 Risk Areas for Chlorophyll a and Faecal Coliform

A total of 107 sampling stations recorded *Chlorophyll a* concentrations during the time period. These sampling stations are primarily located close to the coastal area downstream of urban built-up land cover or cultivated areas. The *Chlorophyll a* risk for the WMA is medium risk as most sample stations measured tolerable concentrations. A total of 2 stations measured acceptable-, 92 tolerable- and 13 unacceptable concentrations of *Chlorophyll a* in the WMA in terms of domestic use (Fig. 10.17).

The St Lucia estuary is of concern as all sampling stations within the vicinity measured tolerable concentrations of *Chlorophyll a*. Sampling stations which measured

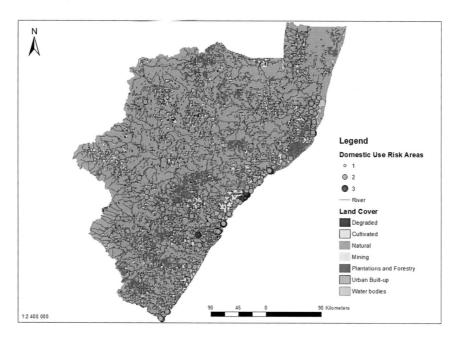

Fig. 10.17 Overall risk profile of *Chlorophyll a* for the Pongola-Mtamvuna WMA (domestic use standards)

unacceptable concentrations are predominantly located downstream of cultivated areas, WWTWs or urban built-up land cover areas. In terms of recreational water quality, the WMA is also mostly of medium to low risk as most sampling stations measured tolerable to acceptable concentrations of *Chlorophyll a*. A total of 66 sample stations measured tolerable, 25 acceptable and 13 unacceptable concentrations (Fig. 10.18).

Same trend exists as in the case of domestic use water quality where most high-risk areas are located downstream of cultivated areas as well as WWTWs and urban built-up. The monitoring network, however, needs to be expanded upon in the WMA as large areas are not being monitored for *Chlorophyll a*. The coastal region of the WMA clearly has significant issues with *Chlorophyll a* which can be mainly attributed to poor or inadequate WWTWs. For the WMA to avoid future eutrophication issues, it will have to invest in the improvement or upgrading of wastewater facilities as these are currently lacking and may exacerbate eutrophication issues in future.

A total of 150 sampling stations measured *Faecal coliform* in the WMA for the time period. In terms of domestic use standards the WMA is of high risk as 114 stations measured unacceptable- and 11 tolerable levels of *Faecal coliform*. Only 25 stations measured acceptable levels of *Faecal coliform* and are mostly located in undeveloped regions downstream of cultivated land cover areas (Fig. 10.19). The whole coastal area, as well as some urban centres located inland of the WMA, are plagued by unacceptable levels. Concerningly, mostly all sampling stations located

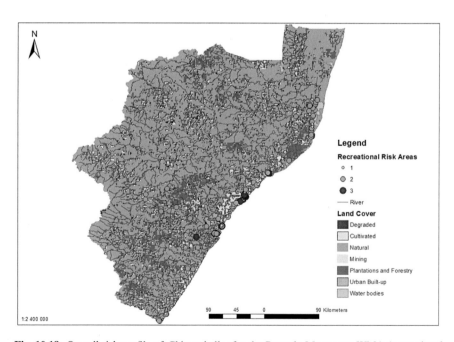

Fig. 10.18 Overall risk profile of *Chlorophyll a* for the Pongola-Mtamvuna WMA (recreational use standards)

10.2 Pongola-Mtamvuna WMA

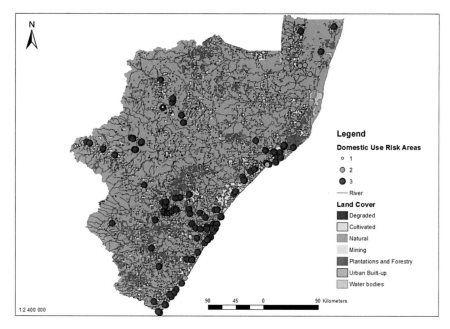

Fig. 10.19 Overall risk profile of *Faecal coliform* for the Pongola-Mtamvuna WMA (domestic use standards)

close to the coastal areas have measured unacceptable levels. Populations living in these areas, therefore, have to be cautious when making use of water from these areas and the development of desalinisation plants needs to take this into account as water obtained from the ocean will be contaminated and will require further treatment to remove *Faecal coliform* to be of an acceptable standard.

In terms of irrigation standards, the situation is a bit better, however, follows the same trend as above (Fig. 10.20). The WMA is predominantly of high to medium risk as most stations recorded unacceptable to tolerable levels of *Faecal coliform* for the period. Only 9 stations recorded acceptable levels and are once again located in undeveloped areas of the WMA.

The water quality of surface water resources is mostly fair except in terms of *Chlorophyll a* and *Faecal coliform*. Additional and more extensive water quality monitoring is however required to be able to fully understand the water quality of the WMA. Large parts of the WMA are in good ecological condition as majority of its rivers are classified as being largely natural to moderately modified. Rivers which are located within the vicinity of urbanised areas are classified as largely to seriously modified and are immensely impacted by urban built-up land use and associated activities. Most of the WMAs water resources are not stressed or under great threat. Water resources of rivers which have been largely or significantly modified are however stressed and under threat and requires precautionary approaches in the management thereof to maintain acceptable conditions. The WMA does, however,

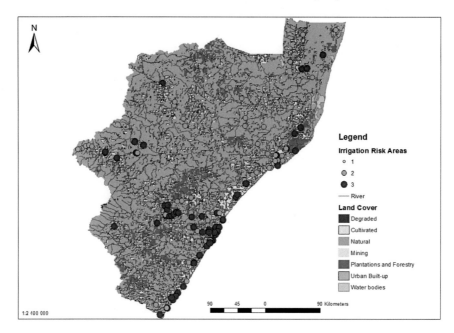

Fig. 10.20 Overall risk profile of *Faecal coliform* for the Pongola-Mtamvuna WMA (irrigation standards)

have to invest in the expansion of the water quality network but more importantly in the improvement or upgrading of current WWTWs as most of these facilities are not up to standard. The water resources of the WMA may become more stressed due to eutrophication problems caused by unacceptable levels of *Chlorophyll a* and may also face human health problems due to unacceptable levels of *Faecal coliform* especially in coastal regions.

10.3 Orange WMA

10.3.1 WMA Overview

The Orange WMA is of critical importance to South Africa and is made up of the Lower and Upper Orange catchments (Fig. 10.21). The Vaal River system is augmented from the Upper Orange (Senqu) by the Lesotho Highlands Water Project which supplies the economic heartland of the country. Thermal power stations located in the Highveld, irrigation schemes along the Vaal, Middle and Lower Orange rivers is also supplied by the river system. The Orange basin supplies approximately 15 million people which are heavily dependent on it.

10.3 Orange WMA

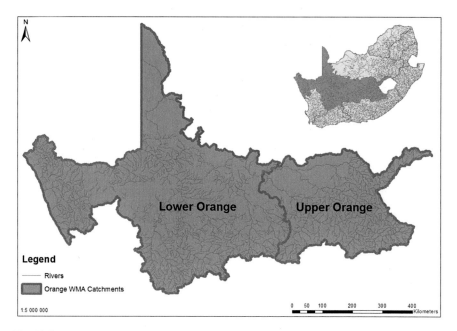

Fig. 10.21 Main catchments of the Orange WMA

The Orange River originates in the Drakensberg Mountains in Lesotho and flows towards the west through South Africa to the Atlantic Ocean at Alexander Bay. It is consequently the longest river in the country (2,200 km) with a total basin area of 973,000 km^2. The major tributaries within the WMA include the Vaal, Modder, Riet, Kraai and Caledon. The main storage dams in the Orange River are Gariep and Vanderkloof, Welbedacht Dam in the Caledon River, Rustfontein, Mockes, and Krugersdrift Dams in the Modder River with the Tierpoort and Kalkfontein Dams in the Riet River.

The Upper River stretches from the origin of the Senqu River to its confluence with the Vaal River. The Upper Orange catchment's land use is dominated by natural vegetation with the main economic activity being livestock farming. Large areas under dryland cultivation focussed mainly on grain production, primarily located in the north-east of the catchment. The Modder Riet sub-catchment is primarily focussed on agricultural activities with limited mining and few urban centres. Ficksburg is primarily known for cherry orchards in the region. Large areas under irrigation are primarily for growing of grain and fodder crops, located along the main rivers downstream of irrigation dams. Mangaung (previously Bloemfontein), Botshabelo and Thaba 'Nchu are the main urban and industrial developments and two large hydropower stations have been developed at the Gariep and Vanderkloof Dam. Mining has declined overall in the catchment and current activities mainly relate to salt works and small diamond mining operations.

The Lower Orange catchment includes the Orange River between the Orange-Vaal confluence and Alexander Bay. The Orange River forms a green strip in an arid region and also forms the border between South Africa and Namibia. Tributaries include the Ongers and Hertebeest rivers from the South and the Molopo River and Namibia Fish River from the North. The catchment is characterised by highly intermittent water courses along the coast which drain into the ocean. The Lower Orange is the largest catchment but is the driest and most sparsely populated catchment in the country. Minerals and water from the Orange River is the key for economic development in the region. Irrigation is the dominant water use sector in the Lower Orange, using 94% of the total water requirements. The importance of agriculture is attributed to the climate which is suitable for the growth of high-value crops, together with water availability along the Orange River.

The Orange River's flow regime and water quality has severely been impacted by extensive upstream developments. Salinity in the river has also increased due to the transfer of good quality water from the Orange River in Lesotho and Upper Orange WMA as well as due to saline irrigation return flows along the Orange River and its main tributaries. The poor quality water from the Vaal River, containing high proportion of irrigation return flows, mining drainage as well as treated urban effluent also periodically enter the Orange River, significantly negatively affecting its water quality. Current water demands on the Orange System is generally in balance with supply but additional demand will have to be met by increasing supply through building more storage or improving management of existing uses through water demand management and conservation strategies.

As indicated previously, the Orange WMA is dominated by natural landcover. The degraded land use cover is mostly bare soil areas (Fig. 10.22). Cultivated areas mostly occur in the north-eastern region of the WMA.

Water quality monitoring of surface water resources is limited especially within the Lower Orange catchment. Low water monitoring sites can be attributed to the catchment being mostly underdeveloped and being characterised by highly intermittent watercourses. A total of 88 sampling stations had suitable data for the period and are mostly located in the Upper Orange catchment. Of these 88 sample stations, 18 are dams/barrages-, 42 rivers-, 2 spring/eyes- and 26 WWTWs sampling stations.

10.3 Orange WMA

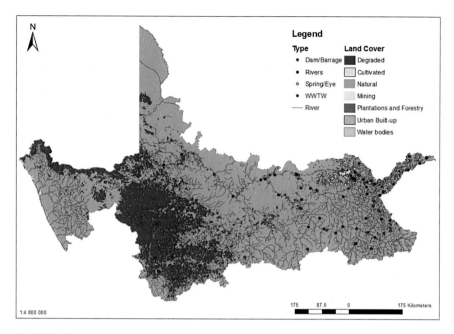

Fig. 10.22 Water quality sampling sites used and land cover of the Orange WMA

10.3.2 Risk Areas for Domestic Use

The Orange WMA is dominated by no-risk areas, followed by medium-risk areas (Fig. 10.23). A total of 50 (57%) sampling stations are classified as no-risk areas and 20 (23%) medium-risk areas. Only seven stations are classified as high risk.

Medium-risk areas are spread over the WMA especially within the Upper Orange catchment and are primarily located downstream of cultivated areas as well as urban built-up and WWTWs. No medium or high-risk areas were recorded in the Lower Orange catchment. Of the 20 medium-risk areas, only 1 is a dam/barrage sampling station and 2 spring/eyes. The other 17 medium-risk areas are all located within close proximity or directly downstream of WWTWs. The 7 high-risk areas which were established are all WWTWs sampling stations located in the Upper Orange catchment. The WWTWs within this catchment should, therefore, receive attention as most are not up to standard. The functioning and standard of effluent from these facilities should, therefore, be investigated to establish which steps need to be taken for these facilities to achieve acceptable standards of the selected water quality parameters.

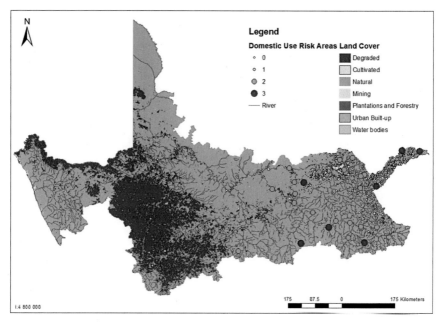

Fig. 10.23 Overall risk profile and significant risk areas for the Orange WMA (domestic use standards)

10.3.3 Risk Areas for Aquatic Ecosystems

The aquatic ecosystem water quality for the WMA is predominantly of low to medium risk. Of the 88 sampling stations, 57 sampling stations are of low risk, 23 of medium risk and 8 of high risk. Once again, these risk areas are located in the Upper Orange catchment especially close to cultivated and urban built-up land cover (Fig. 10.24).

Medium-risk areas are primarily located in the eastern half of the WMA and include 1 dam/barrage-, 4 rivers-, 1 spring/eye and 17 WWTWs sampling stations. High-risk areas are found close to cultivated and urban built-up land cover areas and are all WWTWs sampling stations. To ensure that the WMAs aquatic ecosystems are not degraded further, focus has to be placed on WWTWs as these facilities are clearly functioning below standard and are lacking proper management.

10.3.4 Risk Areas for Irrigation Use

The WMAs water quality for irrigation use is predominantly of no risk (Fig. 10.25). Of the 88 sampling stations, 52 (59%) are of no risk, 11 (12%) low risk, 18 (20%) medium risk and 7 (9%) of high risk all located in the Upper Orange catchment.

10.3 Orange WMA

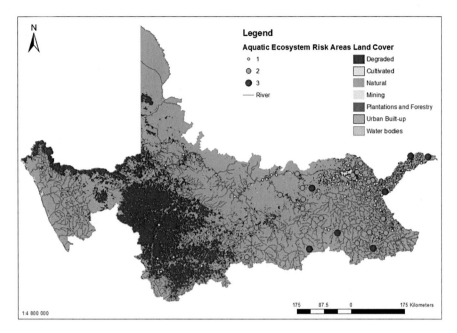

Fig. 10.24 Overall risk profile and significant risk areas for the Orange WMA (aquatic ecosystem standards)

Medium-risk areas to similar to those identified for domestic use and aquatic ecosystem water quality standards. Of the 18 identified medium-risk areas, 1 dam/barrage and 2 spring/eye sampling stations were classified as medium-risk areas. The other 15 stations are all WWTWs. In terms of high-risk areas, all 7 sampling stations are WWTWs. Farmers or irrigation schemes making use of water resources close to these areas or facilities should, therefore, reserve caution as high-risk areas are characterised by tolerable to unacceptable standards of pH, EC, chloride, sodium, ammonia and nitrate concentrations.

10.3.5 Risk Areas for Industrial Use

Industrial water quality for the Orange WMA is predominantly of low to medium risk. A total of 49 (56%) sampling stations were classified as low risk and 29 (33%) as medium-risk areas. Only 2 sampling stations were classified as being of no risk (Fig. 10.26).

Medium and high-risk areas are located in both the Upper and Lower Orange catchments. Medium-risk areas include 2 river sampling stations in the Lower Orange catchment and 1 dam/barrage-, 5 river-, 2 spring/eye- and 19 WWTWs sampling stations located in the Upper Orange catchment. High-risk areas according to industrial

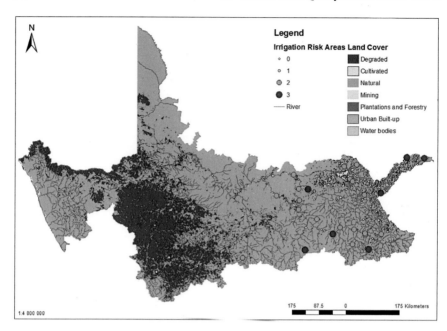

Fig. 10.25 Overall risk profile and significant risk areas for the Orange WMA (irrigation standards)

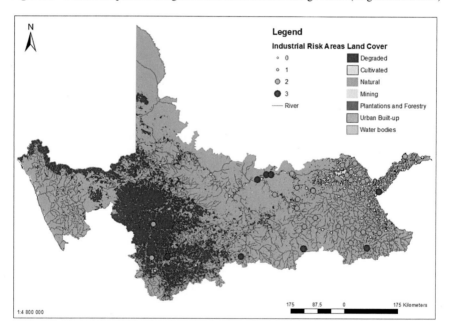

Fig. 10.26 Overall risk profile and significant risk areas for the Orange WMA (industrial standards)

10.3 Orange WMA

water quality guidelines are primarily found in the Lower Orange catchment with 2 river- and 2 dam/barrage sampling stations recording mostly tolerable to unacceptable standards of pH, EC, chloride and sulphate. High-risk areas for the Upper Orange catchment include 1 river- and 3 WWTWs. All WWTWs sampling stations located in the Orange WMA recorded tolerable to unacceptable water quality standards for industrial water use in terms of the selected water quality parameters. Industries located downstream of these WWTWs facilities need to take note as the use of these water resources may affect the efficiency of operations and lead to unintended financial expenses due to the poor quality of water resources.

10.3.6 Risk Areas for Chlorophyll a *and* Faecal Coliform

The Orange WMA only has a total of 11 sampling stations which measure *Chlorophyll a* concentrations and are located along the course of the Orange River and on some tributaries located in the Upper Orange catchment. In terms of domestic use water quality standards, the WMA is of medium to high risk as nine of the 11 sampling stations measured tolerable standards. Unacceptable standards were recorded by 2 sampling stations either located downstream of urban built-up, specifically WWTWs, or downstream of agricultural activities within the Upper Orange catchment (Fig. 10.27).

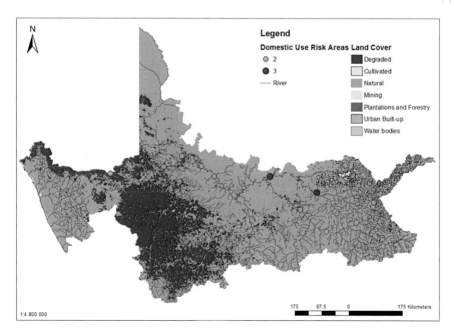

Fig. 10.27 Overall risk profile of *Chlorophyll a* for the Orange WMA (domestic use standards)

In terms of recreational water quality standards, the WMA is predominantly of medium risk as 6 sampling stations measured tolerable concentrations. Only 3 stations measured acceptable concentrations of *Chlorophyll a* and are located in natural land cover areas. High-risk areas are the same as above with 1 high-risk sample station located downstream of a WWTWs and other downstream of cultivated land cover (Fig. 10.28).

A total of 112 sampling stations recorded levels of *Faecal coliform* within the Orange WMA. In terms of domestic use water quality standards, the WMA is of high risk as 111 (99%) sampling stations recorded unacceptable levels of *Faecal coliform* which should be of great concern. Most sampling stations are located throughout the Upper Orange catchment with 3 sampling stations located in the Lower Orange catchment (Fig. 10.29).

The Orange WMA is of medium to high risk in terms of recreational water quality standards as 62 (55%) of sampling stations recorded tolerable levels and the other 50 (45%) sampling stations unacceptable levels (Fig. 10.30). High-risk areas are mostly located within close proximity or directly downstream of cultivated land cover or urban built-up. The overall tolerable to unacceptable levels of *Faecal coliform* in the WMA can be attributed to animal waste connected to livestock farming activities as well as below average effluent and wastewater from urban built-up areas, specifically WWTWs. The WWTWs facilities in the WMA can, therefore, be described as to be below average to poor as water quality measured downstream of these facilities are

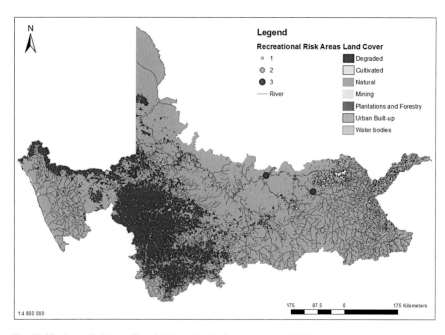

Fig. 10.28 Overall risk profile of *Chlorophyll a* for the Orange WMA (recreational use standards)

10.3 Orange WMA

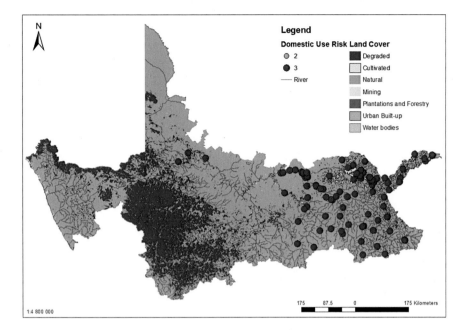

Fig. 10.29 Overall risk profile of *Faecal coliform* for the Orange WMA (domestic use standards)

predominantly of tolerable to unacceptable standards for all of the selected water quality parameters.

The water monitoring network within the WMA is therefore limited to the main stem of the Orange River with monitoring frequency being very intermittent. The Upper Orange is of an acceptable level in terms of salinity due to most of the water flowing from the Highlands of Lesotho into the Senqu River. Salinity however worsens becoming of a tolerable standard in the middle of the Lower Orange River. The Modder and Riet rivers are of a tolerable to unacceptable state primarily due to the impact of irrigation return flows as well as urbanisation and below average wastewater treatment. Salinity is of an unacceptable range at the Douglas Barrage on the Vaal River (just upstream of the confluence with the Orange River). The poor status of the barrage can be attributed to the impact of upstream irrigation activities including those from the Modder Riet catchment.

The ecological state of the WMA varies. The Upper Orange River is moderately to largely modified. Ecological state improves to moderately modified to largely natural state from the Augrabies to the Orange River Mouth. Smaller tributaries are predominantly of a moderately modified state and largely modified state with a very small percentage of tributaries in less developed areas being of a largely natural state.

Half of the WMAs catchments can be described as to be stressed surface water resources which are under threat which will require precautionary approaches to management to maintain good conditions. The WMAs water resources do still have

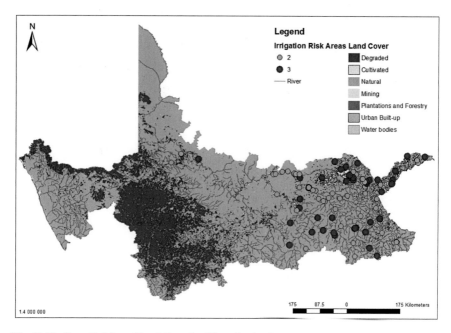

Fig. 10.30 Overall risk profile of *Faecal coliform* for the Orange WMA (irrigation standards)

the capacity to accept degrees of impact, however, this capacity will start to dwindle if medium-risk areas are not addressed.

10.4 Conclusions

The central WMAs vary significantly according to the established risk areas. All three of the WMA are mostly of low risk however significant risks were established which may have widespread environmental effects as well as possibly be accompanied by significant human health risks especially in terms of general unacceptable levels of *Faecal coliform*.

The Vaal WMA is characterised by both formal and informal urbanisation, industrial growth, agricultural activities and widespread mining activities. The Upper Vaal WMA is largely developed and plays a significant role in the country's economy as it serves the Gauteng Province with water. Significant risks were identified for most of the evaluated water quality standards in terms of the selected physical and chemical water quality parameters. The WMA is predominantly of low risk in terms of most of the water quality guidelines for the selected physical and chemical parameters. The Middle Vaal was found to be of highest risk as multiple significant risk areas were established which can be attributed to high concentration of cultivated areas as well as urban built-up and mining operations. WWTWs were found to be the biggest

10.4 Conclusions

culprit for tolerable to unacceptable concentrations of EC, sodium, ammonia, nitrate and phosphate throughout the WMA.

The Vaal WMA had very few sampling stations which measured *Chlorophyll a* concentrations and should be addressed. Most sampling stations measured tolerable to unacceptable levels of *Chlorophyll a* which can be attributed to runoff from cultivated areas and urban built-up areas as well as discharge of below standard wastewater from WWTWs. The Vaal WMA, especially in terms of the Upper Vaal catchment has an extensive *Faecal coliform* problem as almost all sampling stations recorded unacceptable levels. Unacceptable levels of *Faecal coliform* can mainly be attributed to runoff of animal wastes from cultivated land cover, inadequate sewage treatment facilities for scattered small towns and rural settlements as well as urban centres. WWTWs needs to be addressed throughout the WMA as this currently poses significant human health risks and will contribute to continued environmental degradation especially in terms of aquatic ecosystems which are already under stress due to anthropogenic activities.

The Pongola-Mtamvuna WMA is facing a major challenge in supplying additional water supply needed to meet the growing needs of the Kwazulu-Natal Coastal Metropolitan Area which includes Durban-Pietermaritzburg, KwaDukuza in the North to Amanzimtoti in the South. Water requirements which are constantly increasing and catchments are already in deficit. The WMA is predominantly of low to no risk for most of the selected water quality parameters in terms of physical and chemical water quality parameters. The *Chlorophyll a* monitoring network is mainly limited to the coastal region of the WMA, downstream of urban centres. *Chlorophyll a* concentrations were found to be mostly of a tolerable standard. Tolerable concentrations can be attributed to runoff of nutrients or wastewaters from cultivated and urban built-up areas. *Faecal coliform* is a major issue as most of the WMA is characterised by unacceptable to tolerable levels. Coastal regions especially those located downstream of urban centres or rural settlements are especially of high risk. The WMAs population, therefore, need to serve caution when making use of water for domestic or recreational use as it poses significant human health risk. The constructed desalinisation plants which will make use of these coastal waters also need to be aware of the major *Faecal coliform* risk as additional treatment will be required to ensure that water is suitable for domestic use.

The Orange WMA is also of critical importance to the country as the Orange River is the key for economic development in the region. The WMA supplies water to 15 million people and is predominantly focused upon agricultural activities in the Upper Orange catchment. The Lower Orange catchment is characterised by bare and dryland cover and is sparsely populated. The WMA is predominantly of low to no risk for most of the selected water quality parameters in terms of physical and chemical water quality parameters. The *Chlorophyll a* monitoring network is very limited as it only had 11 sampling stations which measured the parameter. The limited number of sampling stations measured tolerable to unacceptable levels of *Chlorophyll a* and is mainly attributed to runoff from cultivated areas. In terms of *Faecal coliform*, the WMA showed the same trend as the Vaal and Pongola-Mtamvuna WMAs. The WMA is challenged by unacceptable to tolerable levels of *Faecal coliform* mostly in the more developed Upper Orange catchment. This is once again mainly due to inadequate or non-functioning WWTWs as well as runoff of animal wastes from cultivated areas as well as the absence of proper sewage treatment facilities for rural settlements.

The central WMAs need to expand their water quality monitoring network especially in terms of the measurement of *Chlorophyll a*, as eutrophication is an identified major water quality problem in the country. Monitoring of *Faecal coliform* should also be expanded as it is a major issue. Significant risk areas once again directly correlate with the highly altered nature of the catchments. Therefore, areas which are characterised by anthropogenic activities need to have more detailed monitoring in terms of all water quality parameters to ensure the proper identification of water quality problems before levels become of a tolerable or unacceptable nature.

All three of the WMAs also need to place a significant focus on the improvement of their WWTWs are these facilities are either inadequate, ill-maintained or not functioning properly due to mismanagement. Proper sewage treatment facilities also need to be constructed for rural settlements. These significant risk areas need to be addressed to minimise or limit future environmental degradation, significant human health risks as well as socio-economic costs.

Chapter 11
Current Water Quality Risk Areas for Berg-Olifants, Breede-Gouritz and Mzimvubu-Tsitsikamma WMAs

The focus is placed on the WMAs located in the southern region of the country namely the Berg-Olifants, Breede-Gouritz and the Mzimvubu-Tsitsikamma WMAs. All of the WMAs are predominantly of low to no risk, however, numerous significant risk areas were established for all of the WMAs and directly correlated with the extent of modification of water sources or areas.

Significant risk areas were predominantly established downstream or within close proximity of urban centres, and cultivated areas with most high and medium risk areas being directly downstream or close proximity of WWTWs. WWTWs are of great concern for all of the WMAs as it is dominated by unacceptable to tolerable levels of most or all selected water quality parameters especially in terms of *Faecal coliform*. Most of the WWTWs facilities within these WMAs do not comply with set standards and can be attributed to these facilities being mismanaged, inadequate or in need of proper maintenance or upgrading. Proper sewage facilities are also needed to be developed for rural settlements. The WMAs could invest in the reuse of wastewater after it has invested in the upgrading or maintenance of current WWTWs as the reuse of wastewater could be seen as an untapped water source which could lessen pressure on catchments. Desalinisation of ocean water needs to be accompanied with proper treatment due to *Faecal coliform* being a major issue. Desalinised water will have to go through proper treatment to ensure water is of a proper domestic use standard and eliminate any microbiological threats such as *Faecal coliform*. These issues need to be addressed to avoid future significant environmental human health problems and risks and to ensure constant future water supply.

11.1 Berg-Olifants WMA

11.1.1 WMA Overview

The Berg-Olifants WMA, located in the south-western region of South Africa, includes the Berg and Olifants–Doorn catchments and major rivers, Berg, Diep, Steenbras, Olifants, Doorn, Krom, Sand and Sout (Fig. 11.1). The WMA is situated in the Western Cape and Northern Cape Provinces which includes Cape Town, the second most populous metropolitan area in the country. Several large towns are also located in the WMA with economies based on tourism, education, agriculture and industry. Natural vegetation and soil areas dominate the WMA. Large natural vegetation areas comprise of Cape Fynbos which represents one of the unique floral kingdoms in the world and is recognised as a World Heritage Site. Consequently, a lot of conservation and heritage sites are found in the WMA. The WMA is also characterised by large spatial variations of rainfall, water availability, level and nature of economic development, population density and the potential for development for growth.

The Berg River catchment is made up of the Upper Berg area which includes the Berg River catchment down to the Misverstand Weir located in the Lower Berg area which includes downstream reaches of the Berg River and endoreic areas along the west coast, including the Diep River catchment and the Greater Cape Town Area in the southern area of the WMA. The Olifants–Doorn catchment is made up of the Olifants and Doring River sub-catchments which are characterised by well-watered valleys of the Olifants River sub-catchment, the arid Doring River sub-catchment as well as the highly developed Sandveld area which forms the western coast boundary.

The Berg River catchment's economy is diversified and is mainly dominated by industrial and other activities of the Cape Town Metropolitan area. Other significant economic sectors include irrigated agriculture (wine production), table grapes and deciduous fruit exports as well as tourism. The Olifants–Doorn catchment, on the other hand, is dominated by extensive commercial agriculture which includes irrigated citrus, deciduous fruits, grapes and potatoes. Other economic activities in the catchment include tourism, livestock farming, some industries related to food processing and packaging and limited forestry.

It is important to note that water resources are fully developed in the WMA and investigations are underway to assess other options to augment water supply which will include water conservation and demand management, infrastructure development, reuse of water, groundwater exploitation as well as desalinisation.

11.1 Berg-Olifants WMA

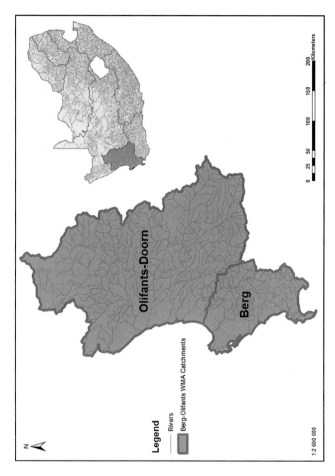

Fig. 11.1 Main catchments of the Berg-Olifants WMA

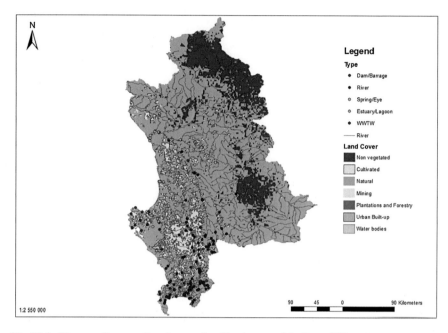

Fig. 11.2 Water quality sampling sites used and land cover of the Berg-Olifants WMA

The surface water quality monitoring network is primarily focused in the southern area of the WMA within the Cape Town Metropolitan area and close neighbouring areas (Fig. 11.2). A total of 181 sampling stations were evaluated which included 5 dams/barrages, 91 rivers, 12 spring/eyes, 17 estuary/lagoons and 56 WWTWs.

11.1.2 Risk Areas for Domestic Use

The surface water quality of the Berg-Olifants WMA in terms of domestic use standards is a combination of no risk in some parts as well as low to medium risk in others. A total of 71 (39%) sampling stations were classified as no risk areas, while 49 (27%) are low risk and 44 (24%) medium risk. High-risk areas include 17 sampling stations (Fig. 11.3).

Significant risk areas are primarily located around urban centres such as the Cape Town Metropolitan area in the south as well as around and downstream of cultivated areas in the Berg catchment. Medium-risk areas are spread across the WMA but primarily located in the Berg River catchment (south). A total of 41 sampling stations are of medium risk in the Berg River catchment and include 11 river-, 1 spring/eye- and 29 WWTWs sampling stations. In terms of high-risk areas, only 2 rivers were found to be of tolerable to unacceptable standards, however, 14 WWTWs were found to be of high risk. The WWTWs in the Berg River catchment are, therefore, of

11.1 Berg-Olifants WMA

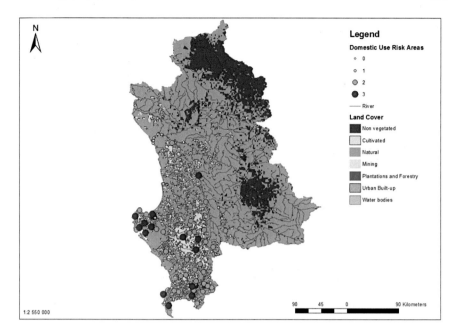

Fig. 11.3 Overall risk profile and significant risk areas for the Berg-Olifants WMA (domestic use standards)

medium to high risk as most sampling points located directly downstream measured tolerable to unacceptable concentrations of the selected physical and chemical water quality parameters. The same trend exists in the Olifants–Doorn catchment where 3 of the 4 WWTWs are of medium risk and 1 of high risk. Populations located close or downstream of these facilities, therefore, need to be cautious when using water directly from rivers as tolerable to unacceptable concentrations were measured for EC, calcium, chloride, sodium, ammonia, nitrate and phosphate.

11.1.3 Risk Areas for Aquatic Ecosystems

The water quality standards in terms of aquatic ecosystems guidelines largely vary across the WMA. Of the 181 stations, 44 (24%) sampling stations were classified as no risk, 62 (34%) as low risk, (17%) 30 as medium risk and (25%) 45 as high risk (Fig. 11.4).

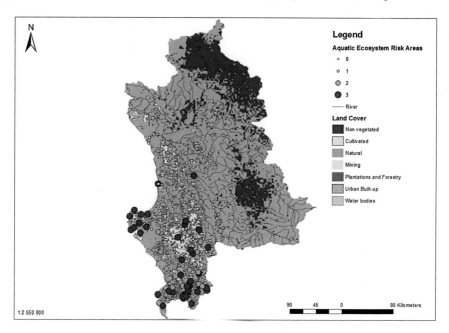

Fig. 11.4 Overall risk profile and significant risk areas for the Berg-Olifants WMA (aquatic ecosystem standards)

Most medium- and high-risk areas are located close or directly downstream of cultivated and urban built-up land cover areas. The Berg River catchment is once again of higher risk than the Olifants–Doorn catchment. Medium-risk areas in the Berg River catchment include 11 river-, 1 spring/eye- and 15 WWTWs sampling points and high-risk areas include 7 river- and 36 WWTWs sampling points. In terms of the Olifants–Doorn catchment, 3 of the 4 WWTWs are of medium risk, 1 WWTWs high risk and 1 river sampling point of high risk. WWTWs are, therefore, of major concern as it is the primary risk area within this WMA and should be addressed if the degradation of aquatic ecosystems is to be decreased or minimised.

11.1.4 Risk Areas for Irrigation Use

The irrigation water quality for the WMA is predominantly of no to low risk. A total of 76 (42%) sampling stations were classified as no risk areas and 59 (33%) as low risk (Fig. 11.5).

A total of 33 sampling stations were classified as medium risk and 13 as high risk. Most of these areas of significant risk are mostly located in the Berg River catchment. The Berg River catchment contains 11 river-, 1 spring/eye- and 18 WWTWs sampling stations classified as medium-risk areas and 12 WWTWs as high-risk areas. The same trend exists in the Olifants–Doorn catchment where once again 3 of the 4 WWTWs are of medium risk and the other of high risk. Cultivated areas located close or downstream of these WWTWs, therefore, need to serve caution as tolerable to unacceptable levels of EC, calcium, chloride, sodium, ammonia, nitrate and phosphate were recorded.

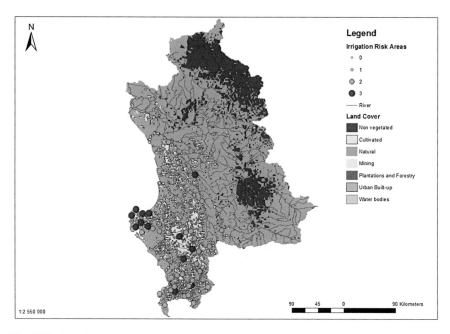

Fig. 11.5 Overall risk profile and significant risk areas for the Berg-Olifants WMA (irrigation standards)

11.1.5 Risk Areas for Industrial Use

The Berg-Olifants WMA predominantly of low risk as 106 (59%) sampling stations recorded mostly acceptable standards for the selected physical and chemical water quality parameters. A total of 35 (19%) sampling stations are classified as no risk, 27 (15%) medium risk and 13 (8%) of high risk (Fig. 11.6).

Once again, the Berg River catchment contains most of the medium-high-risk areas. Medium-risk areas in the Berg River catchment were identified as 8 river-, 1 spring/eye and 11 WWTWs and high-risk areas 9 river and 2 spring/eyes sampling points. In terms of the Olifants–Doorn catchment, medium-risk areas include 1 river-, 3 spring/eye- and 3 WWTWs sampling points and only 2 high-risk areas which include 1 river- and 1 WWTWs sampling site.

Significant risk areas for industrial water use are primarily located downstream of cultivated areas or urban built-up specifically downstream of WWTWs. Tolerable to unacceptable concentrations of pH EC, chloride and sulphate were recorded by these sampling stations. Current industries located within close proximity of these areas and which use water directly for processes as well as planned industries close to these locations need to take note of these medium- and high-risk areas as poor water quality may affect processes and lead to unintended financial losses through decreased productivity.

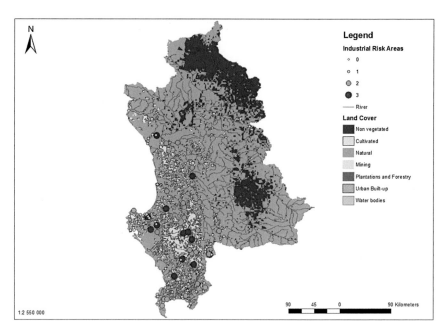

Fig. 11.6 Overall risk profile and significant risk areas for the Berg-Olifants WMA (industrial standards)

11.1.6 Risk Areas for Chlorophyll a *and* Faecal Coliform

A total of only 4 sampling stations recorded *Chlorophyll a* in the whole WMA and a proper evaluation can therefore not be completed. The WMA lacks *Chlorophyll a* water quality monitoring and should, therefore, invest in the development and expansion of these sampling stations. In terms of domestic use water quality guidelines, all 4 of the sampling stations recorded tolerable concentrations of *Chlorophyll a* (Fig. 11.7).

In terms of recreational standards, 2 stations recorded acceptable standards and 2 recorded tolerable *Chlorophyll a* concentrations all located close or downstream of cultivated land cover areas (Fig. 11.8).

In terms of the monitoring of *Faecal coliform*, a total of 127 sampling stations measured the parameter in the WMA. These sampling stations are spread across the WMA, mainly located in the Berg River catchment close to the Cape Town Metropolitan area. The WMA is mostly of unacceptable standard in terms of domestic use as 125 sampling stations recorded unacceptable levels of *Faecal coliform* and only 2 tolerable (Fig. 11.9). Most of these sampling stations are located within close proximity of urban built-up areas as well as cultivated land cover. Populations should, therefore, not use water directly from these areas as it may lead to negative health effects. In terms of irrigation standards, the WMA measured predominantly unacceptable levels of *Faecal coliform* (84 or 66% of sample stations) to tolerable (43

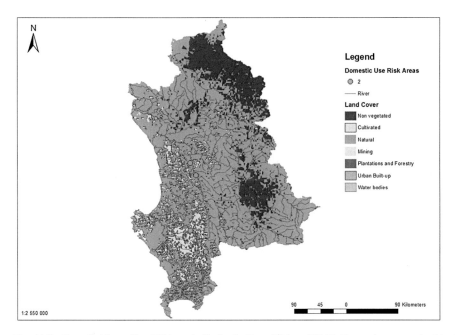

Fig. 11.7 Overall risk profile of *Chlorophyll a* for the Berg-Olifants WMA (domestic use standards)

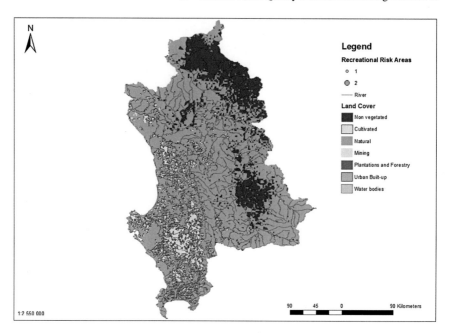

Fig. 11.8 Overall risk profile of *Chlorophyll a* for the Berg-Olifants WMA (recreational use standards)

or 34%). Once again most of the high-risk areas are located close to built-up areas, however, some high-risk areas are located downstream or close to some cultivated areas in the WMA (Fig. 11.10). Farmers should, therefore, take note of these risk areas when using this water directly from sources for irrigation purposes as water is not of an acceptable standard in terms of *Faecal coliform* levels.

The limited spatial distribution of water quality monitoring stations, especially in terms of sampling stations which measure *Chlorophyll a* concentrations, make it difficult to complete an evaluation of water quality for certain areas of the WMA. The lower reaches of the Olifants–Doorn river is, however, of an unacceptable standard in terms of salinity. The state of the Berg River is currently in the acceptable range in the upper reaches, however, deteriorates to unacceptable levels downstream which is mainly attributed to agricultural activities and anthropogenic impacts.

The Berg River catchment is predominantly in a moderately modified state and largely modified in the southern and western parts of the WMA. Most of the Olifants–Doorn catchment is in a natural to a largely natural state with a small percentage of the river reaches in the vicinity of urban centres and agricultural areas which have consequently been degraded and described as seriously modified. Most of the WMA's rivers are in a modified condition primarily due to agricultural activities and urban development. Half of the WMA's surface water resources are, therefore, stressed and under threat and require precautionary approaches to manage and maintain present good or acceptable conditions. The WMA's surface water resources

11.1 Berg-Olifants WMA   257

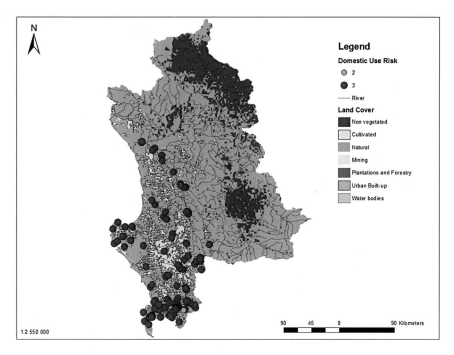

Fig. 11.9 Overall risk profile of *Faecal coliform* for the Berg-Olifants WMA (domestic use standards)

also have the capacity to absorb degrees of impact, however, if the current state of WWTWs are not addressed as well as if the various sources of pollution and associated effects of agricultural and urban developments are not minimised, this capacity will dwindle and lead to widespread environmental as well as socio-economic effects across the WMA especially in the south and western parts.

11.2 Breede-Gouritz WMA

11.2.1 WMA Overview

The Breede-Gouritz WMA is made up of the Breede, Overberg, the Karoo and Klein Karoo and Outeniqua Coastal Area (Stilbaai to Plettenberg Bay) catchments. Major catchments include the Breede and Gouritz catchments (Fig. 11.11).

It is located in the south-western region of the country, lies predominantly within the Western Cape Province with small portions in the Eastern and Northern Cape Provinces. Major rivers include the Breede, Sonderend, Sout, Bot, Palmiet, Gouritz,

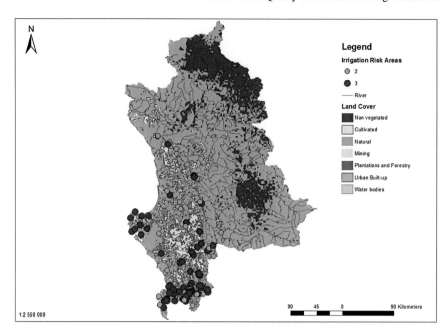

Fig. 11.10 Overall risk profile of *Faecal coliform* for the Berg-Olifants WMA (irrigation standards)

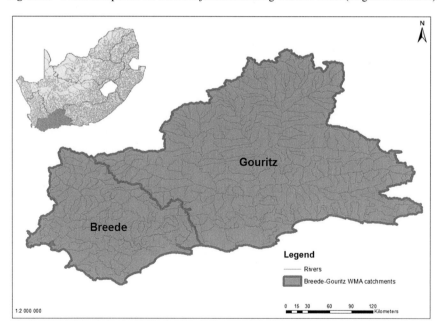

Fig. 11.11 Main catchments of the Breede-Gouritz WMA

Olifants, Kamanassie, Gamka, Buffels, Touws, Goukou and Duiwenhoks. Most of the WMA is of a rural nature.

The Breede-Overberg catchment is mainly made up of mountain ranges, the Breede River valley as well as the hills of the Overberg in the south. The Breede River is extensively utilised with two large dams which include the Brandvlei and Theewaterskloof. The catchment also contains numerous small dams as well as a large number of farm dams. The major towns within the catchment include Worcester and Ceres with numerous other small towns which include Grabouw, De Doorns, Robertson, Swellendam, Montagu, Caledon, Hermanus and Gansbaai. Groundwater is of great importance in the catchment as it supplies many towns and farms. It is important to note that the Breede catchment is becoming more saline due to intensive irrigation in the area and have impacted the surface water quality negatively.

The Palmiet River is also intensively farmed. The lower reaches of the river is protected as part of the Kogelberg Biodiversity Reserve which requires that the ecological condition is maintained. The Klein Karoo catchment is vast and dry with some water flowing through the Swartberg mountain range (situated in the north of the WMA). This area includes Beaufort West as the major town in the north-west area which is largely reliant on groundwater. Other smaller towns include Oudtshoorn as well as De Rust in the Gouritz catchment area which also includes the Dwyka, Groot, Gamka and Olifants tributaries. The Gouritz River is the main river in this catchment, contributing to a large proportion of surface flow. The catchment is made up of good arable land, however, irrigation is limited due to low and variable rainfall.

Existing resources have been over-allocated with no opportunity for further dam development. The Klein Karoo catchment is also water stressed as there is no additional surface water development available to support the growing needs of towns such as Oudtshoorn, Dysseldorp and other surrounding areas. Potential does, however, exist to exploit groundwater to try and augment water supply. Elevated salinity is a natural occurrence within the inland catchments of the Karoo and Klein Karoo due to the geology of the area. Impacts on the region's water quality have been observed largely due to land-based activities in the more populated areas.

The Outeniqua coastal sub-catchment which stretches from Stilbaai to Outeniqua, is ecologically sensitive containing many short steep rivers considered of high ecological importance. The sub-catchment, therefore, contains a number of National Parks and conservation areas. The catchment is, however, also a major growth area as it is a popular retirement location and year-round tourist destination. It also includes small- to medium-sized dams which include the Wolwedans and Garden Route Dams. Its surface water resources are almost fully developed and alternative supplies are, therefore, required.

Land use in this area is characterised by commercial agriculture in the Breede and Overberg areas. Commercial agricultural activities include irrigated agriculture which includes wine and table grapes, dairy and deciduous fruit, livestock farming, dry land agriculture which includes wheat and canola, as well as other associated activities such as packaging and processing. These mentioned activities form the basis the WMA's economy. It also produces 70% of South Africa's table grapes, apples and fynbos export. Tourism and residential development along the coast also play a role as key economic drivers in the region. The agricultural sector in the Karoo to Klein Karoo and Outeniqua coastal areas, provide the primary economic driver for the region by producing a large variety of crops, livestock and fruit. The fish and shellfish industry, tourism as well as the ostrich industry also play significant roles in the economy of the coastal region and land use is consequently dominated by irrigation and afforestation activities.

Water quality monitoring of surface water resources is primarily located on major rivers and tributaries, close to land-based activities as well as urban built-up areas (Fig. 11.12). Large parts of the central region have no water quality monitoring points which can mainly be attributed to land cover being of a bare, non-vegetated nature not supporting land-based activities. A total of 249 sampling stations had suitable data for the period. Of these 249 sampling stations, 29 are dam/barrages-, 140 rivers-, 10 spring/eyes-, 31 estuary/lagoon-, 1 wetland- and 38 WWTWs sampling stations.

11.2.2 Risk Areas for Domestic Use

The Breede-Gouritz WMA is predominantly of no risk in terms of domestic use water quality. A total of 122 (49%) sampling stations were classified as no risk and are mostly located in underdeveloped areas. Low-risk areas include 63 (25%) sampling

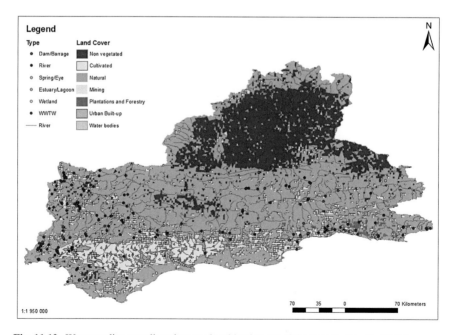

Fig. 11.12 Water quality sampling sites used and land cover of the Breede-Gouritz WMA

11.2 Breede-Gouritz WMA

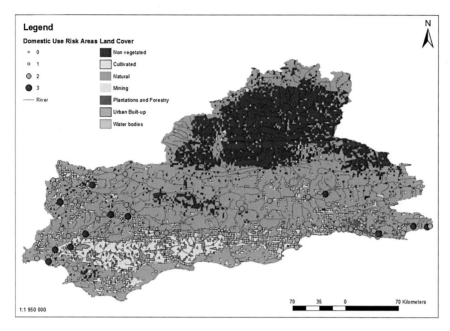

Fig. 11.13 Overall risk profile and significant risk areas for the Breede-Gouritz WMA (domestic use standards)

stations, medium risk 52 (21%) and 12 (5%) sampling stations were classified as high risk (Fig. 11.13). Medium- and high-risk areas are spread across the WMA.

Most risk significant risk areas are, however, located in the Gouritz catchment which includes 6 river-, 3 dam/barrage-, 15 estuary/lagoon and 7 WWTWs sampling stations of medium risk and 4 WWTWs sampling stations of high risk. All WWTWs sampling stations recorded tolerable to unacceptable concentrations of the selected water quality parameters which should be of great concern. In terms of the Breede catchment, 5 river-, 2 estuary/lagoon- and 14 WWTWs were classified as medium risk and 8 WWTWs as high risk. Most of the catchment's WWTWs are also characterised by tolerable to unacceptable concentrations primarily of EC, calcium, chloride, sodium, ammonia, nitrate as well as phosphate.

11.2.3 Risk Areas for Aquatic Ecosystems

The Breede-Gouritz WMA is predominantly of low to medium risk in terms of aquatic ecosystem standards. A total of 150 (60%) sampling stations were classified as low risk and 60 (24%) of no risk (Fig. 11.14). Significant risk areas are once again spread across both main catchments, however, in this case the Breede catchment recorded more risk areas in terms of aquatic ecosystem water quality standards. The

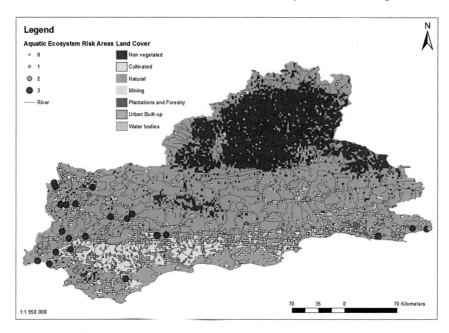

Fig. 11.14 Overall risk profile and significant risk areas for the Breede-Gouritz WMA (aquatic ecosystem standards)

Breede catchment contains 3 river- and 4 WWTWs sampling points of high risk and 20 WWTWs sampling points of high risk. In the case of the Gouritz catchment 1 dam/barrage- and 8 WWTWs sampling points were classified as medium risk and 3 WWTWs sampling points as high risk.

Most river and dam/barrages which have been classified as medium or high risk are predominantly located downstream of cultivated land cover areas or close to urban areas. Tolerable to unacceptable concentrations can, therefore, be attributed to agricultural and urban runoff.

11.2.4 Risk Areas for Irrigation Use

In terms of irrigation water quality standards, the WMA is predominantly a mixture of no, low and medium risk. A total of 98 (39%) sampling stations were classified as no risk, 76 (31%) as low-, 50 (20%) as medium- and 25 (10%) as high risk. Most medium risk areas are located within the Gouritz catchment (Fig. 11.15). Medium risk areas in the catchment include 9 river-, 3 dams/barrage-, 15 estuary/lagoon- and 6 WWTWs sampling stations and 3 WWTWs sampling stations as high risk. In terms of the Breede catchment, the catchment contains most of the high-risk areas. Medium risk areas in the catchment include 12 rivers- and 5 WWTWs sampling

11.2 Breede-Gouritz WMA

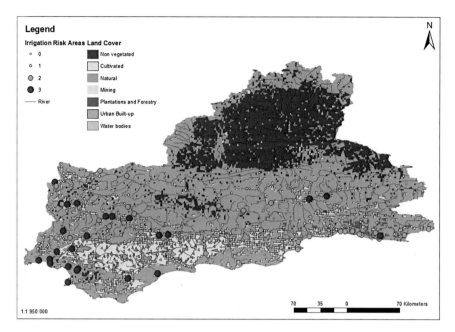

Fig. 11.15 Overall risk profile and significant risk areas for the Breede-Gouritz WMA (irrigation standards)

stations and high-risk areas 2 river-, 1 wetland, 2 estuary/lagoon- and 17 WWTWs sampling stations. Most of the WMA's WWTWs are once again characterised as being medium to high risk areas.

Cultivated land cover areas, making use of irrigation water at these locations, within close proximity or downstream of these facilities need to reserve caution as most sampling stations classified as high risk recorded tolerable to unacceptable concentrations of EC, ammonia, nitrate as well as phosphate.

11.2.5 Risk Areas for Industrial Use

The Breede-Gouritz WMA is predominantly characterised by low risk in terms of industrial water quality. A total of 137 (55%) sampling stations recorded mostly acceptable levels while 36 (14%) sampling stations were classified as no risk. Medium-risk areas include (22%) 54 sampling stations and high risk of 22 (9%) (Fig. 11.16).

The Gouritz catchment is predominantly of medium risk and the Breede catchment of higher risk. Medium-risk areas located in the Gouritz catchment include 10 river-, 6 dam/barrage-, 1 spring/eye-, 14 estuary/lagoon- and 8 WWTWs sampling stations.

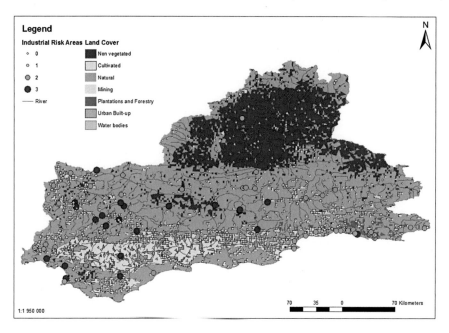

Fig. 11.16 Overall risk profile and significant risk areas for the Breede-Gouritz WMA (industrial standards)

High-risk areas were recorded as 6 river-, 1 dam/barrage- and 1 estuary/lagoon sampling stations.

The Breede catchment contains 9 river-, 1 wetland- and 5 WWTWs sampling stations as well as 10 river-, 2 spring/eye- and 2 estuary/lagoon sampling stations. Most high-risk rivers are located downstream of cultivated land cover areas and measured tolerable to unacceptable concentrations of EC and chloride.

Most WWTWs have been classified as medium to high risk and measured tolerable to unacceptable concentrations of pH, EC as well as chloride. Industries which are located downstream of these facilities and use water directly from the water resources, therefore, need to serve caution as unacceptable concentrations of EC may have negative consequences for production processes or equipment in the long term.

11.2.6 Risk Areas for Chlorophyll a and Faecal Coliform

A total of only 19 sampling stations measured *Chlorophyll a* in the WMA. The number of sampling stations are therefore not enough to establish the overall risk for the WMA and investments need to be made in expanding this water quality monitoring network. Of the 19 sampling stations, 4 measured acceptable standards and 15 measured tolerable standards in terms of domestic use water quality (Fig. 11.17).

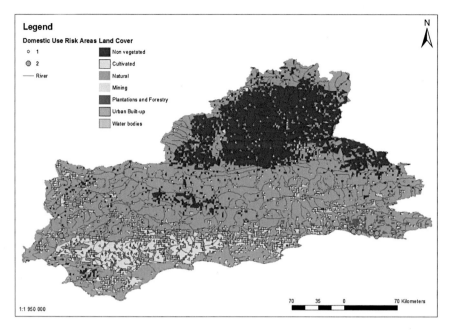

Fig. 11.17 Overall risk profile of *Chlorophyll a* for the Breede-Gouritz WMA (domestic use standards)

Most sampling stations which measured tolerable concentrations are either located downstream of cultivated areas as in the case in the western region of the WMA or downstream of afforestation areas on the eastern half of the WMA close to the coastal area. The runoff from these land cover areas, therefore, contribute to mostly tolerable concentrations of *Chlorophyll a* in terms of domestic use guidelines.

In terms of recreational standards, the WMA is predominantly of an acceptable standard as 16 sampling stations measured acceptable concentrations while only 3 measured tolerable concentrations for the period (Fig. 11.18). The tolerable sampling stations are once again located downstream of cultivated and afforestation land cover areas. Parts of the estuary located in the Breede catchment has, however, measured tolerable concentrations in terms of both domestic use and recreational standards and should be noted to ensure that the estuary is not degraded further or to ensure that degradation is minimised.

The water quality monitoring for *Faecal coliform* in the WMA is a bit more extensive than in the case of *Chlorophyll a* as a total of 118 sampling stations recorded its levels over the period. The WMA is of an unacceptable standard in terms of domestic use water quality guidelines as 115 sampling stations recorded unacceptable levels of *Faecal coliform*, 1 sampling station tolerable levels and only 2 sampling stations acceptable levels (Fig. 11.19). In terms of irrigation water quality, the WMA is predominantly of tolerable to unacceptable *Faecal coliform* levels and spread

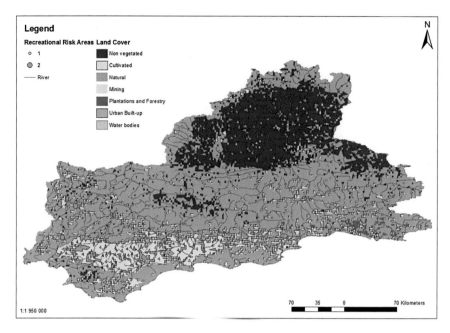

Fig. 11.18 Overall risk profile of *Chlorophyll a* for the Breede-Gouritz WMA (recreational use standards)

across the WMA (Fig. 11.20). A total of 64 sampling stations recorded tolerable levels and 54 sampling stations unacceptable levels for the period.

Unacceptable levels of *Faecal coliform* are primarily recorded downstream of cultivated land cover areas which can be attributed to runoff of animal wastes as well as downstream or within urban built-up areas and WWTWs. The WMA should, therefore, place focus on addressing current WWTWs as these facilities are in need for upgrading if found to be inadequate and/or require improved maintenance. Great attention needs to be given to these facilities especially from a human health perspective as *Faecal coliform* is associated with significant human health effects. It is, therefore, advised that the region's population do not make use of untreated water from various water resources as most of these water resources have unacceptable to tolerable levels of *Faecal coliform*.

The overall water quality status of the WMA can be described as to be acceptable to tolerable, however, the water quality monitoring network needs to be expanded upon. The current ecological condition in the Breede to the Gouritz side of the WMA is mainly in a moderately modified state and largely modified to the southern and western parts of the WMA. Smaller tributaries in the Karoo catchment (north) are primarily in a natural to largely natural state as these are located in less developed and impacted areas. Some river reaches in the vicinity of urban areas/towns and agricultural areas have been severely degraded and are classified as seriously modified. Most

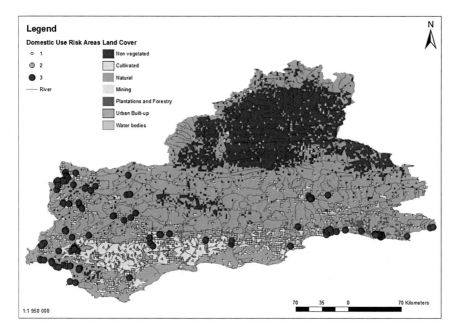

Fig. 11.19 Overall risk profile of *Faecal coliform* for the Breede-Gouritz WMA (domestic use standards)

of the rivers in the WMA are in modified conditions due to impacts from agricultural activities and urban development as established in the preceding sections.

The WMA is therefore characterised by some degree of water-stressed surface water resources which are under threat. A very small percentage requires precautionary approaches to management to maintain good condition and most surface water resources have the capacity to accept degrees of impact. With continued urban development as well as the continued rundown of current inadequate WWTWs, the WMA will experience an increase in water stress and surface water resources will become more vulnerable to degrees of impact. It is therefore imperative that the WMA's CMA invests in the expansion of its water quality network to identify additional risk areas as well as focus on the upgrading, maintenance or construction of additional WWTWs in the WMA to avoid further degradation and possible human health risks.

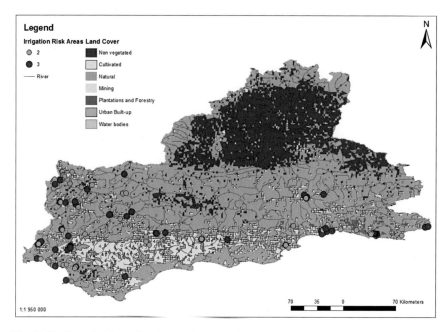

Fig. 11.20 Overall risk profile of *Faecal coliform* for the Breede-Gouritz WMA (irrigation standards)

11.3 Mzimvubu-Tsitsikamma WMA

11.3.1 WMA Overview

The Mzimvubu-Tsitsikamma WMA is made up of the whole Eastern Cape Province and includes large rivers as well as many diverse catchments from the arid Karoo in the west to the subtropical in the north-east. The WMA comprises of main catchments namely the Mzimvubu–Keiskamma and Fish Tsitsikamma catchments which are comprised of the Mzimbuvu, Mtata, Mbashe, Groot Kei, Nahoon, Buffalo, Keiskamma, Boesmans River, Great Fish, Sundays, Kowie, Kromme, Groot, Gamtoos and Tsitsikamma sub-catchment areas which rivers drain directly into the Indian Ocean (Fig. 11.21).

The Mzimvubu River is the largest undeveloped river in the country and flows through deep gorges across the coastal plain before it discharges into the Indian Ocean at Port St. Johns. The Amatola coastal catchment contains main rivers of the Buffalo, Keiskamma and Nahoon which drain in a south-easterly direction into the ocean near the city of East London. The Great Kei catchment drains the northern slopes of the Amatola mountain range and the southern slopes of the Stormberg/Drakensberg range. The Great Kei River exits into the Indian Ocean at the Kei Mouth situated north of East London. Catchments associated with Great Fish and Sundays River

11.3 Mzimvubu-Tsitsikamma WMA

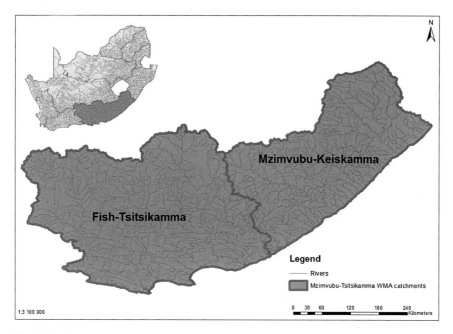

Fig. 11.21 The main catchments of the Mzimvubu-Tsitsikamma WMA

extend from the watershed of the Orange River system to the shoreline of the east coast of the country. The Fish and Sundays catchments are, however, very dry.

The Krom River drains the narrow valley between the Sundays Mountains in the interior and the Tsitsikamma Mountains at the coast. The Gamtoos River catchment includes the Groot and Kouga Rivers as its major tributaries. The Groot River catchment lies in the Karoo and the Kouga River originates in the Baviaanskloof Valley. The Groot and Kouga Rivers join to form the Gamtoos River which drains the western slopes of the Elandsberg mountain range to the Indian Ocean.

The climate and temperature variations found in the catchment is closely related to the elevation and proximity to the coast. Areas along the coast experience a mild, temperate climate with more extreme conditions inland with summer rainfall.

The main urban areas in the WMA include Nelson Mandela Bay (Port Elizabeth, Uitenhage and Despatch) and Buffalo City and the towns of Grahamstown, Craddock and Queenstown. A large percentage of the WMA's population are located in rural areas with their incomes being dependent on the agricultural sector, which is mainly subsistence farming. Extensive irrigation agriculture occurs alongside the Fish and Sundays Rivers. Other dominant economic activities include tourism, commercial forestry as well as manufacturing with vehicle manufacturing being the dominant industry in the Buffalo City Municipal Areas. Significant future growth is expected for the Buffalo City Municipal and the Nelson Mandela Bay Municipal areas due to employment opportunities attracting populations from smaller urban centres and rural areas.

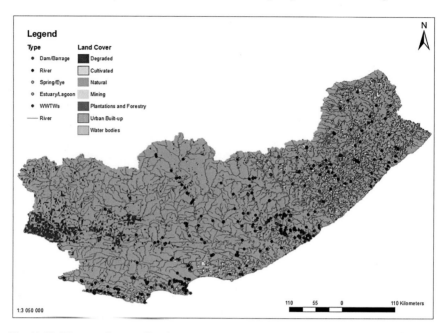

Fig. 11.22 Water quality sampling sites used and land cover of the Mzimvubu-Tsitsikamma WMA

The Mzimvubu to Keiskamma area can be described as water rich in which water resources have not yet been fully developed. Small hydroelectric developments are present in the area and an inter-basin transfer occurs between the Kei and Mbashe catchments. The water requirements in the area are therefore much less than the potential yield and this trend is likely to continue. There are few areas which exceed the local water resources yield and where interventions are needed. The development of future dams is being looked at to develop additional water supply for Queenstown, Buffalo City Municipality, Albany Coast and towns in the Bushman's River catchment.

Future water resource development and interventions are, however, required for the Nelson Mandela Metropolitan Municipality to support growth in water requirements. Water requirements of the Great Fish and Sunday catchments are being met through water transfers from the Orange River. The development of groundwater as well as the improved management thereof has been identified as a necessity to be able to meet water requirements and support different water services in multiple areas of the WMA.

The water quality monitoring of surface water resources in the WMA is concentrated on major rivers and their tributaries, along the coastal area as well as in the vicinity of urban areas. A total of 380 sampling stations had suitable data for the period and were evaluated. Of the 380 sampling stations, 30 are dam/barrage-, 240 river-, 6 spring/eye-, 1 estuary/lagoon- and 103 WWTWs sampling stations (Fig. 11.22).

11.3.2 Risk Areas for Domestic Use

The Mzimvubu-Tsitsikamma WMA is characterised by no to medium risk areas in terms of domestic use water quality guidelines. A total of 184 (48%) sampling stations are classified as no risk, 61 (16%) as low-, 100 (26%) as medium- and 35 (10%) as high-risk areas. Significant risk areas are spread across the WMA (Fig. 11.23). The Fish Tsitsikamma catchment was established to have a total of 59 medium risk areas and 24 high-risk areas. Sampling points which were classified as medium risk include 27 rivers, 2 spring/eyes and 30 WWTWs and high-risk areas include 4 rivers, 1 estuary/lagoon and 19 WWTWs. Most high-risk areas are located close to the coastal area, downstream of urban built-up or WWTWs facilities.

In terms of the Mzimvubu–Keiskamma catchment covering the eastern half of the WMA, 41 medium- and 11 high-risk areas were established. These significant risk areas include 10 river- and 31 WWTWs sampling points of medium risk and 1 river and 10 WWTWs sampling points of high risk. Most of these risk areas are located downstream or within close vicinity of major urban centres. Mostly tolerable to unacceptable concentrations were recorded for EC, calcium, chloride, sodium, ammonia, nitrate and phosphate were recorded at these risk areas and untreated water should therefore not be used for domestic use within these areas.

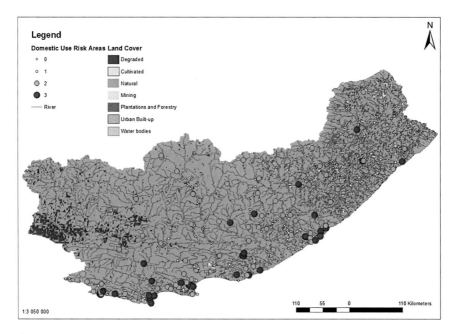

Fig. 11.23 Overall risk profile and significant risk areas for the Mzimvubu-Tsitsikamma WMA (domestic use standards)

11.3.3 Risk Areas for Aquatic Ecosystems

The WMA is dominated by low-risk areas in terms of aquatic ecosystem water quality guidelines (Fig. 11.24). A total of 179 sampling stations were classified as low risk and 83 as low risk areas. Medium-risk areas include 84 sampling stations and high risk 34. Medium risk areas are spread across the WMA while high-risk areas are predominantly located on the coastal region as well as within the vicinity of urban centres. The Fish Tsitsikamma catchment once again hold the largest overall risk. Medium risk areas include 4 river-, 1 estuary/lagoon- and 19 WWTWs sampling points and 15 river-, 1 spring/eye- and 27 WWTWs sampling points. The Mzimvubu–Keiskamma catchment is characterised by 10 river- and 31 WWTWs sampling points of medium risk and 1 river- and 10 WWTWs sampling points of high risk.

Most risk areas are closely associated with WWTWs and urban centres. Current operations of WWTWs, therefore, need to receive attention as effluent from these facilities are not up to standard. Tolerable to unacceptable standards of chloride, ammonia, nitrate and phosphate were recorded at all high-risk areas. To limit or minimise possible further degradation of aquatic ecosystems, WWTWs either need to be upgraded and/or better maintained.

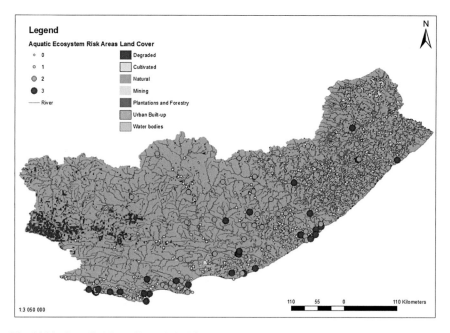

Fig. 11.24 Overall risk profile and significant risk areas for the Mzimvubu-Tsitsikamma WMA (aquatic ecosystem standards)

11.3.4 Risk Areas for Irrigation Use

Irrigation water quality within the WMA is mostly of no risk. A total of 174 (46%) sampling stations were classified as no risk, 89 (23%) low risk, 82 (22%) medium risk and 35 (9%) high risk (Fig. 11.25).

Medium-risk areas are spread across the WMA whereas most high-risk areas are located along the coastal region. The Fish Tsitsikamma catchment is mostly of medium risk and include 27 river-, 1 dam/barrage-, 2 spring/eye-, 1 estuary/lagoon- and 27 WWTWs sampling stations. High-risk areas can be found at 4 river-, 1 spring/eye- and 19 WWTWs. In terms of the Mzimvubu–Keiskamma catchment, it is also dominated by medium-risk areas which include 6 river- and 20 WWTWs- and high-risk areas, 2 river and 9 WWTWs sampling stations. These sampling stations predominantly measured tolerable to unacceptable concentrations of EC, chloride, sodium, ammonia as well as nitrate over the time period. Irrigated cultivated areas using water directly from water resources located within the vicinity of these identified risk areas need to be aware of these possible high concentrations as it may affect crop growth and production.

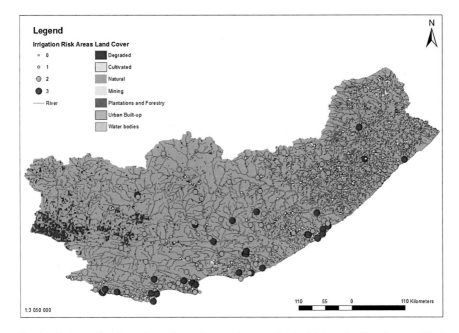

Fig. 11.25 Overall risk profile and significant risk areas for the Mzimvubu-Tsitsikamma WMA (irrigation standards)

11.3.5 Risk Areas for Industrial Use

Industrial water quality within the WMA is predominantly of low to no risk. A total of 192 (51%) sampling stations were classified as low risk, 86 (23%) as no risk, 77 (20%) as medium risk and 25 (6%) as high risk (Fig. 11.26).

The Fish Tsitsikamma catchment holds a bit more risk than the Mzimvubu–Keiskamma catchment. A total of 18 river-, 1 dam/barrage-, 2 spring/eye- and 13 WWTWs sampling points are of medium risk within the Fish Tsitsikamma catchment. High-risk areas in the catchment include 12 river-, 1 dam/barrage-, 2 spring/eye-, 1 estuary/lagoon and 2 WWTWs sampling points. The Mzimvubu–Keiskamma catchment was evaluated to have 16 river-, 5 dam/barrage- and 27 WWTWs sampling points of medium risk and 2 river- and 5 WWTWs sampling points of high risk.

It is advised that industrial activities located downstream or within close proximity of these high-risk areas not use untreated water from the applicable water resource as tolerable to unacceptable concentrations of pH, EC and chloride were measured.

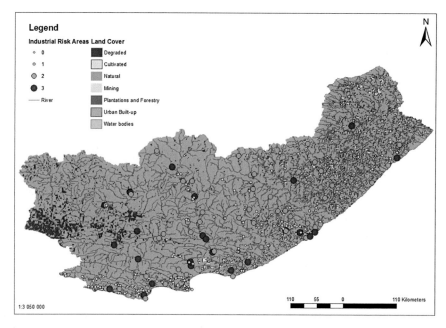

Fig. 11.26 Overall risk profile and significant risk areas for the Mzimvubu-Tsitsikamma WMA (industrial standards)

11.3.6 Risk Areas for Chlorophyll a *and* Faecal Coliform

The Mzimvubu-Tsitsikamma WMA only had 60 sampling stations which recorded *Chlorophyll a* concentrations over the period and are predominantly located close to the Nelson Mandela Metropolitan area. A holistic evaluation for the WMA as a whole can therefore not be provided due to the limited *Chlorophyll a* water quality monitoring network.

Regions monitored are characterised by tolerable levels of *Chlorophyll a* followed by some low- and high-risk areas in terms of domestic use water quality guidelines (Fig. 11.27). In terms of the recreational water quality guidelines, the monitored regions are characterised by tolerable to acceptable levels (Fig. 11.28).

The *Faecal coliform* water quality network is more extensive as in the case of *Chlorophyll a* within the WMA. A total of 317 sampling stations recorded levels of *Faecal coliform* over the period. The WMA is predominantly of an unacceptable standard in terms of domestic use water quality guidelines as 308 sampling stations recorded unacceptable levels of *Faecal coliform* and is spread across the WMA (Fig. 11.29). It is, therefore, advised that the region's population do not use untreated water for domestic purposes as the WMA's does not comply.

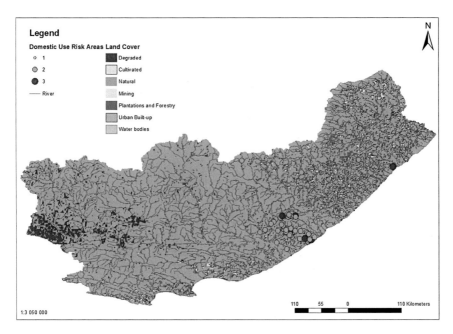

Fig. 11.27 Overall risk profile of *Chlorophyll a* for the Mzimvubu-Tsitsikamma WMA (domestic use standards)

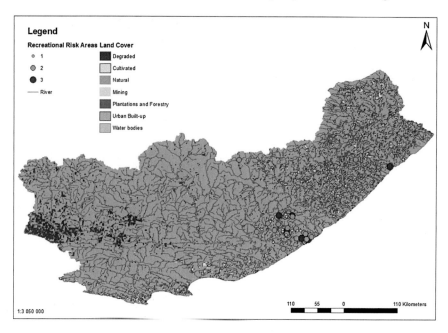

Fig. 11.28 Overall risk profile of *Chlorophyll a* for the Mzimvubu-Tsitsikamma WMA (recreational use standards)

The *Faecal coliform* levels in terms of irrigation water quality standards for the WMA is from an unacceptable to tolerable level as 180 (57%) sampling stations measured unacceptable levels and 133 (42%) sampling stations tolerable (Fig. 11.30). Most sampling stations which measured tolerable or unacceptable levels are located within the vicinity of cultivated and urban built-up land cover. Below standard levels of *Faecal coliform* can, therefore, be attributed to runoff containing animal waste from these cultivated areas or deposition of untreated sewage from urban built-up areas. Farms using these water resources for irrigation purposes should, therefore, take note as it may have unintended human health consequences.

In conclusion, the water quality in the WMA especially in terms of *Chlorophyll a* is not well monitored and limited to reaches of the Mzimvubu catchment, upper tributaries of the Kei River catchment and then selected quaternary catchments in the Fish and Sundays River catchment areas. Improved monitoring is, therefore, required.

Current water quality varies across the catchment with good water quality found in the upper tributaries, becoming of a worse standard especially at coastal areas. Salinity is a major issue within coastal areas which is attributed to the saline nature of rivers due to the area's geology as well as due to intensive irrigation return flow.

The ecological conditions of the Mzimvubu catchment area are predominantly good as most of the catchment is in a natural to moderately modified state. The Groot Kei and Amatole regions are mainly in a moderately to largely modified state

11.3 Mzimvubu-Tsitsikamma WMA

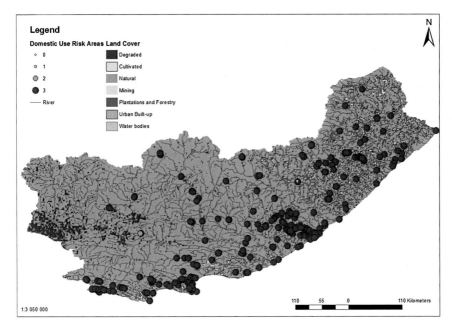

Fig. 11.29 Overall risk profile of *Faecal coliform* for the Mzimvubu-Tsitsikamma WMA (domestic use standards)

with its smaller tributaries in less developed areas being of a natural to largely natural ecological condition. The Fish and Tsitsikamma catchments are predominantly of a largely natural to moderately modified state but rivers located in the vicinity of Cradock, Port Elizabeth, Uitenhage and other smaller towns are largely modified. Most of the WMA's surface water resources are not under stressed conditions or under major threat and will be able to accept certain degrees of impact on future development in the catchment.

11.4 Conclusions

The southern WMAs vary significantly according to the established risk areas.

The Berg-Olifants WMA includes Cape Town, the second most populous metropolitan area in the country, as well as several large towns with economies based on tourism, education, agriculture and industry. Natural vegetation and soil areas, however, dominate the WMA with large natural vegetation areas comprising of the Cape Fynbos which is one of the unique floral kingdoms in the world and is recognised as a World Heritage Site. Significant risks were identified for most of the evaluated water quality standards in terms of the selected physical and chemical water quality parameters. The WMA is predominantly of a low to no risk in terms of

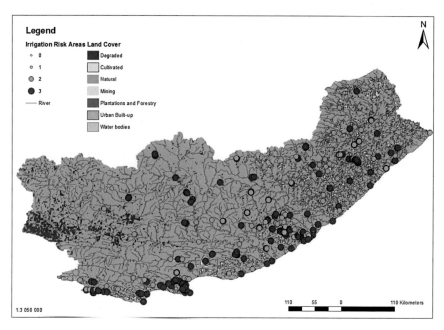

Fig. 11.30 Overall risk profile of *Faecal coliform* for the Mzimvubu-Tsitsikamma WMA (irrigation standards)

most of the water quality guidelines for the selected physical and chemical parameters. The Berg catchment was found to be of highest risk as multiple significant risk areas were established which can be attributed to high concentration of cultivated areas and urban built-up. WWTWs were found to be the biggest culprit for tolerable to unacceptable concentrations of EC, sodium, ammonia, nitrate and phosphate throughout the WMA.

The Berg-Olifants WMA had only 4 sampling stations which measured *Chlorophyll a* concentrations and should be addressed. Most sampling stations measured tolerable levels of *Chlorophyll a* which can be attributed to runoff from cultivated areas and urban built-up areas as well as discharge of below standard wastewater from WWTWs. The Berg-Olifants WMA, especially in terms of the Berg catchment has an extensive *Faecal coliform* problem as almost all sampling stations recorded unacceptable levels. Unacceptable levels of *Faecal coliform* can mainly be attributed to runoff of animal wastes from cultivated land cover, inadequate sewage treatment facilities for scattered small towns and rural settlements as well as urban centres. WWTWs needs to be addressed throughout the WMA as this currently poses significant human health risks and will contribute to continued environmental degradation especially in terms of aquatic ecosystems.

The Breede-Gouritz WMA is mostly of a rural nature and mostly contains small towns and intensive irrigation on cultivated areas. The WMA is predominantly of low to no risk for most of the selected water quality parameters in terms of physical

11.4 Conclusions

and chemical water quality parameters. The *Chlorophyll a* monitoring network is mainly limited certain coastal areas of the WMA at urban centres. *Chlorophyll a* concentrations were found to be mostly of a tolerable standard. Tolerable concentrations can be attributed to runoff of nutrients or wastewaters from cultivated and urban built-up areas. *Faecal coliform* is a major issue as most of the WMA is characterised by tolerable to unacceptable levels. Coastal regions especially those located downstream of urban centres or rural settlements are especially of high risk. The WMA's population, therefore, need to serve caution when making use of water for domestic or recreational use as it poses significant human health risks.

The Mzimvubu-Tsitsikamma WMA is made up of diverse catchments from the arid Karoo in the west to the subtropical in the north-east. The WMA is predominantly of low to no risk for most of the selected water quality parameters in terms of physical and chemical water quality parameters. The *Chlorophyll a* monitoring network is limited as it only had 60 sampling stations which measured the parameter. The limited number of sampling stations measured tolerable to unacceptable levels of *Chlorophyll a* and is mainly attributed to runoff from cultivated and urban built-up areas. In terms of *Faecal coliform* the WMA is challenged by unacceptable to tolerable levels across the whole WMA. This is once again mainly due to inadequate or non-functioning WWTWs as well as runoff of animal wastes from cultivated areas and ultimately the absence of proper sewage treatment facilities for rural settlements.

These WMAs need to expand their water quality monitoring network especially in terms of the measurement of *Chlorophyll a*, as eutrophication is an identified major water quality problem in the country. Monitoring of *Faecal coliform* should also be expanded as it is a major issue. Significant risk areas once again directly correlate with the highly altered nature of the catchment. Therefore, areas which are characterised by anthropogenic activities need to have more detailed monitoring in terms of all water quality parameters to ensure the proper identification of water quality problems before levels become of a tolerable or unacceptable nature.

All three of the WMAs also need to place a significant focus on the improvement of their WWTWs are these facilities are either inadequate, ill-maintained or not functioning properly due to mismanagement. Proper sewage treatment facilities also need to be constructed for rural settlements. These significant risk areas need to be addressed to minimise or limit future environmental degradation, significant human health risks as well as socio-economic costs.

Chapter 12
South Africa's Water Reality: Challenges, Solutions, Actions and a Way Forward

There is a general international consensus that water quality issues worldwide will have progressive and significant impacts or constraints on economic development. The country's National Water Resource Strategy of 2014 acknowledged that the resource is not receiving the attention and status it deserves and wastage, pollution and degradation is widespread. The sustainability of South Africa's freshwater resources has reached a critical point. Real opportunities exist where South Africa can emerge as a leader in the transition into water-smart economies in Africa, and can be achieved through investment into new cost-effective technologies as well as enterprise innovations which all aim to contribute to ensured water security. Emphasis should be placed on having quality wastewater treatment which can be reused and recycled to ultimately try and ensure future sustainability. Decisive steps, however, need to be taken now as the country cannot afford to wait.

12.1 South Africa's Water Reality

From the previous chapters, it is evident that South Africa's water resources are overstretched and under immense strain in terms of quantity and quality. The sustainability of South Africa's freshwater resources has reached a critical point. Multiple rivers, dams/reservoirs as well as coastal lakes have lost their resilience or buffering capacity to adjust to nutrients or sequestrate toxicants. As early as 1979, it was emphasised by Cillie and Odendaal (1979) that by only reducing pollution of the country's limited water resources and through multiple cycle reuse of water and ultimately, the use of desalinisation can a water crisis be averted in the future. An urgent need for immediate responses to water quality problems such as eutrophication, salinisation, acidification and sewage pollution have been called for by multiple scientific studies and reports and have been laid at the door of the government over the past three decades. The following sections will focus the water realities which South Africa is facing.

12.1.1 Persistent Eutrophication

Eutrophication has been a persistent water quality problem within the country. It is estimated that eutrophic conditions exist in one of five of the country's 75 major impoundments and in 18 of 25 major river catchments (de Villiers and Thiart 2007). Approximately, 28% of surface water samples have been recorded as hypertrophic, 33% eutrophic, 37% mesotrophic and only 3% oligotrophic according to samples taken by the National Eutrophication Monitoring Programme in 2013 (DWA 2013). Cyanobacteria which have included Microcystis and Anabaena species have been found in all major impoundments at levels dependent on the trophic state (Scott 1991; van Ginkel et al. 2000; van Ginkel 2004) and frequent, geographically widespread poisonings of domestic and wild animals by cyanobacterial toxins have been recorded (Oberholster et al. 2009). These toxic blooms are a major threat to the supply of safe drinking water to the whole population of South Africa as the bulk of the country's water resources are stored in dams and its primary source of "raw" water.

The country's eutrophication problem arises mainly from the discharge of inadequately treated sewage or wastewater, dumped into its rivers and dams. Even in cases where wastewater is efficiently treated, the concentration of phosphorus is, in the majority of cases, not reduced to levels that will offset problematical eutrophication. The overall failing of sewage treatment plant infrastructure, as well as sustained inattention thereto, has exacerbated the whole situation. Wastewater treatment plants are, however, not the only sources of nutrients. A wide range of non-point sources further complicates the situation. Examples of non-point sources can include runoff-bearing nutrients from fertilisers or animal husbandry operations as well as stormwater from urban areas (Matthews and Bernard 2015).

If current trends persist, an increase in the incidence of toxin-producing cyanobacteria may be the result (Conradie and Barnard 2012). There is no formal reporting system for cyanobacterial blooms in South African freshwaters but the little available data indicate a frequent and widespread seasonal occurrence (Downing and van Ginkel 2004). Data are therefore very limited but the controversial beta-N-methyamino-L-alanine (BMAA) was found in 97% of cyanobacterial culture collection samples isolated from local waters (Downing 2011), and the country has experienced animal deaths from cyanotoxins for many decades (Harding and Paxton 2001).

Eutrophication is, therefore, a widespread threat and has been well known and documented since late 1960. More than 50% of the country's stored water has become increasingly eutrophic and have consequently also increased the cost of treating water to a desirable standard. The severity of eutrophication, as well as occurrence of cyanobacterial blooms, have come at a cost to the country due to water treatment costs, effects on waterside property values, recreational and tourism losses and negative human health impacts from poor water quality including diarrhoea, cholera and waterborne diseases; animal fatalities; poor aquatic ecology negatively affecting ecosystem services; reduced biodiversity and proliferation of invasive species; and the cost of management and control of aquatic macrophytes (Oberholster and Ashton 2008).

The total cost of eutrophication is likely to extend to hundreds of millions of rands per annum, tolerated across all levels of society particularly affecting the livelihoods and health of the poor and vulnerable. The Department of Environmental Affairs has acknowledged inter alia, "Despite the large amount of research that has been carried out on eutrophication in South Africa's water reservoirs and lakes, the collective understanding of the problem remains limited". Further, that "the impact of eutrophication on water sources ultimately results in the reduction of the quality of life of South Africans" (DEA 2012).

12.1.2 Spreading Salinisation

Salinisation of soil and water is a major environmental problem which occurs around the world. It has been recognised as a persistent water quality problem in the country which has both natural and anthropogenic causes. Natural salinisation is primary geological. Anthropogenic causes are multiple in numbers. A wide variety of human activities are associated with the release of salts in short or long term. Point sources of pollution such as discharge of wastewater by industries cause immediate increases in salt concentrations. Non-point sources or diffuse pollution originates from poorly managed urban settlements, land waste disposal as well as mine residue deposits which impact a larger area and poses greater problems.

It leads to a decrease in fertility of landscapes, negatively influences agricultural activities and ultimately degrades water quality to such an extent that it becomes unfit for domestic, recreational, agricultural and/or industrial use. All of these effects are accompanied by immense economic losses as well as water supply issues and should therefore be placed on the government's agenda.

The country's reservoirs receive more than two-thirds of its annual streamflow and these dams, which are often characterised by large surface areas, causes water to evaporate faster than the rate of flowing rivers. Saline levels consequently rise and the higher temperatures associated with climate variability will exacerbate the situation even further by increasing evaporation.

The high costs associated with the measuring waterlogging and salt-affected soils on South African irrigation schemes, as well as inconsistencies in data collection and reporting methods have resulted in incomplete and often contradictory information on the extent and distribution of salt-affected and waterlogged soils. Many rivers such as the Berg River exhibit high salinities (Bugan et al. 2015).

Acid Mine Drainage (AMD) also plays a role in the increase of salinity of the country's water resources as it is highly saline due to its high concentration of sulphates . AMD consequently contributes to salt build-up and together with greater water losses due to evaporation from increased temperatures plus added pathogen-loaded sewage flows result in even higher salinity.

The Vaal River is threatened by rising saline levels and studies which have examined AMD established that total dissolved solids are low in the Vaal and Grootdraai Dam but increases significantly downwards from the Vaal Barrage. High salinity is attributed to AMD flows from tributaries which drain the Witwatersrand goldfield (Casas et al. 1998; Coetzee et al. 2006; Camden-Smith et al. 2015; Turton 2013; Winde 2010a, b). The only manner in which salinity levels can be lowered in the overall system is to raise the freshwater level by introducing clean water from the Lesotho Highlands Water Project as well as reducing the flow or highly saline AMD. The dilution of pollutants with the use of the Lesotho Highlands Water Project is however not sustainable and the causes or sources of pollution in the Vaal WMA have to be addressed.

12.1.3 Mine Water

Mining operations in South Africa continue to play a fundamental role in the development of the country's economic and political landscape. It has, however, brought wealth to some but at the price of causing extensive ecological damage. South Africa is faced by numerous critical environmental challenges which range from land degradation to the destruction of finite resources. AMD is one of the most risky hazards in terms of its implications.

AMD came to the fore in 2002 in the West Rand Basin in the Gauteng Province when flooding from heavy rains issued 20 million litres of AMD a day into the surrounding water supply.

The country is facing a water security dilemma ecologically whilst being driven by a strong mining industry on the economic front, have placed it in a precarious position over the past decade. This precarious position has been driven by continued discharging of acidic water into its water system which has endangered communities as well as ecosystems along the Vaal and Limpopo Rivers. Undue stress on the country's economy and water stressed environment might be a consequence which might also undermine agricultural and industrial sectors. It has been estimated that the current levels of acid deposition in South Africa might exceed the critical load for acidification of acid-sensitive streams. AMD should, therefore, be of a great concern as it will have widespread ramifications in all spheres.

The dilemma is further exacerbated by leakages which discharge approximately 36 million m^3 of AMD a day into the regions water systems and threaten the Cradle of Humankind heritage site. Residential communities which reside in the vicinity, especially those located along the Vaal and Limpopo Rivers, are also threatened by abandoned mines. The most aggravated case is located in the vicinity of Johannesburg and the Witwatersrand.

An Inter-Ministerial Committee was consequently established and the country's Chamber of Mines has prioritised its obligation to assist the government in addressing the AMD issue. The Chamber of Mines stated that it would assist with technological input as well as address AMDs negative societal consequences. The mining sector has, however, accused the government of showing very little to no leadership regarding AMD over the past decade in which they were aware of the significance of the problem. The mining industry depends on the government to create coherent initiatives as these operations are very expensive for the industry. The mining sector believes that these initiatives were not exercised urgently enough and have contributed to the problem.

The government has partially treated and pumped AMD in recent years. This positive step is, however, long overdue and a more complete solution is required to address the issue of AMD. A treatment facility was brought online by the government in 2014 to prevent further AMD decanting or reaching the surface in the Central Rand basin. Most treatment plants have, however, been left wanting of a complete solution as these only neutralise water, leaving high concentrations of sulphate and other salts which may cause acute health effects and make water unusable for domestic use and possibly agricultural use.

Tailings dams have however not received as much attention. Johannesburg has more than 200 of these waste dumps and they contain elevated concentrations of heavy metals such as radioactive uranium like AMD. The government has primarily placed focus on AMD created underground in the West and Central Rand of Johannesburg. The focus should also be placed on the other sources of AMD which include seepage and runoff from these tailings dams.

A Finnish company, Global EcoProcess Services have, however, entered the scene and has developed and patented a process which they claim will revolutionise the treatment of AMD and industrial waste. The company added boric acid to the water treatment process and discovered a manner in which toxic metals can be separated through a precipitation process which produces cleaner water and insoluble metals which can be used as raw materials or for other purposes. This cleaned water is, however, still not suitable for drinking although it removes almost all metal. It still contains high concentrations of sulphate (700–1,400 mg/l) and in South Africa, the maximum limit is 400 mg/l. South Africa has, however, been identified as a primary market as high volumes of waste water from a single mine can reach up to 300,000 m^3 per day.

AMD is, therefore, one of the most costly socio-economic and environmental impacts of mining which can attribute to more water deficits and restrictions if it is not contained. It can ultimately curtail economic activity in six of the nine provinces of the country. The government should, therefore, work together with the established Inter-Ministerial Committee of the country's Chamber of mines and should take responsibility for remedying past neglect if the AMD dilemma is to be averted or its significant effects be minimised or reduced.

12.1.4 Drowning in Sewage

Even though the country has managed to increase the number of households with access to piped water by 46% and improved living conditions since 2005, it has placed severe pressure on limited water resources and water treatment infrastructure which is already ill-maintained.

It is important to note that South Africa's water quality is also dependent on the government's custodial role. The country has a total of 1,085 water treatment plants for potable water. Nearly, one-fourth are in poor condition and unable to deliver drinking water. Additionally, the country has 824 wastewater treatment works for processing sewage effluent and only 60 of these releases clean water. These figures might be under or overestimated from the results established in Chaps. 9–11.

This liquid waste is discharged back into rivers, drawn by water treatment plants and processed for drinking water. Of the total 5.13 billion litres of sewage treated per day, only 16% or 836 million litres per day is treated up to a standard safe enough to be discharged back into rivers and dams. The rest, i.e. 84% is discharged in untreated or partially treated form. This amount has increased dramatically since 1990 when influx control was stopped. Increase in the population, as well as economic expansion, has not been accompanied with the necessary upgrade of sewage and other water services. Rapid urbanisation has also played a major role by increasing the volume of waste going to wastewater treatment plants. Current infrastructure, especially in large metropolitan areas, simply cannot cope. This together with the unlawful discharge of rainwater from gutters into sewers has also increased peak flows during high rainfall periods. Due to sewage effluent being uncontrolled, it has become the country's biggest source of pollution and poses numerous risks especially for the poor and vulnerable.

The reality is that approximately 50,000 L of sewage are discharged into South Africa's rivers every second and the municipal sewage system has largely collapsed. High levels of coliforms choke the natural life in dams and rivers and cause significant human health problems, and in worst cases death. The country's wastewater treatment works are failing nationwide and associated health effects are exacerbated by the country struggling with antibiotic-resistant forms of tuberculosis unevenly distributed with the poor facing the biggest burden.

The failure of wastewater treatment works can also be attributed to the lack of management and training of employees. There are multiple examples where operators have to use handbooks with missing pages to establish the ratio of chlorine and lime in the treatment process. This ratio should be informed by sampling and laboratory results but in most cases, it seems to be estimations. Other cases are reported in the annual Green Drop Report which is produced by the national water and sanitation department. The government issue these reports to shame these facilities into action as it has little jurisdiction over wastewater treatment plants due to these being run by local municipalities. The 2013 Green Drop report noted that only 10% of the total 824 plants released clean water. The other 90% were breaking the law with a third of these rated as needing critical and urgent repair. The Blue and Green Drop reports have also been delayed since 2015 due to budgetary constraints and shortage of staff. These reports have consequently become the subject of the Promotion of Access to Information Act applications, which were lodged in an effort to get the water department to release the full findings. The department indicated that it would complete the draft 2016 report by the end of October 2017 however this has once again been delayed.

There is also a lack of finance from local government to address the problem and the available financial resources do not recognise the investment required to counter potential economic harm. The department describes themselves as the custodians of the country's water and sanitation services but continually fails to release important reports and ultimately fails to address South Africa's immense sewage pollution dilemma.

12.1.5 Poor or Complete Lack of Management or Responsibility

The different state institutions which are mandated to protect the country's water resources and supply the population with sustainable safe water include the DWS, CMAs as well as Water User Associations. Good governance is therefore vital in the whole supply chain to ensure that these mandates are met. Responsible actions from the private sector as well as individual households are also required to assist in good governance. For water governance to be at its most effective efforts need to be coordinated between the government, private sector, non-governmental organisations, communities as well as individuals. Transparent processes with accompanied high levels of shared information and coordinated actions are also necessary. This enables various stakeholders to actively participate in planning and the implementation of water management activities or actions, develop innovations related to climate variability and to overall improve the integration of water management efforts.

This however is not the case in South Africa.

South Africa is faced by major challenges in terms of water management. The overall lack of expertise at all mandated levels, as well as the associated lack of necessary adequate infrastructure especially in rural areas, has consequently left thousands of people without access to suitable drinking water and sanitation. The country's already limited water resources are continually being polluted and municipal treatment systems are failing. Water quality management in the country has therefore been hampered by poor coordination as well as conflicting approaches.

The overall failure to act in a strategic and proactive manner has become the norm over the past two to three decades and all which has changed is that more people have come to the realisation that there are major problems. Even though South Africa is largely dependent on its reservoirs, it lacks reservoir management skills as well as training. A generation has passed since there was limited activity in this field. The country is largely dependent on reservoirs for its very existence, are the 30th most arid country in the world and the responsible government department does not have an active or functioning directorate of reservoir lake management. This is hugely concerning and emphasises the overall failure of the responsible departments which have been mandated to protect and manage the country's valuable water resources to act.

The country's water quality has not been a key focus area of the national government with the focus being placed on job creation, health care, rural development, land reform and the fight against crime and corruption. Even though the responsible department for water quality collects data, the data is not regularly interrogated or routinely published.

The National Aquatic Ecosystem Health Monitoring Programme also does not contain the words "dam, impoundment or reservoir lake", and may consequently cause these components not to be acknowledged in developed rivers. Reports related to water quality such as eutrophication studies have disputed value in the water resource management structure as it appears that most water quality issues or concepts such as eutrophication have almost been entirely ignored. The government has admitted that reservoirs may require more human and financial resources but an appropriate monitoring programme for reservoirs first have to be investigated (Kalyan 2010; Davidson 2011). This acknowledgement clearly shows the complete lack of high-level understanding of the relevance of reservoirs in the country. The absence of a set of developed reservoir management skills and the formulation of effective solutions for individual reservoirs will remain elusive due to continued inaction.

The near-total lack of structures, skills as well as planning which is required to address water quality issues has consequently reduced the overall water availability especially in terms of water resources stored in reservoirs. Residual skills have been ignored for too long and an economic analysis of the country's increasing water crisis concluded that a 1% decrease in water quality could result in the loss of 200,000 jobs, a decrease of 5.6% in per capita disposal income and an increase of 5% in government spending (Plus Economics 2010). This has been seen as an underestimate as the full ramifications of cumulative impacts of poor water quality may increase these numbers.

A sound judgement or intelligent reasoning would dictate that it is better to prevent water resources from becoming contaminated with sewage and wastewater than allow already stressed water resources to be continually degraded, having to treat the resultant consequences and still hope to have suitable water quality for domestic use and other water uses. The country's powers that be needs to incorporate individuals with high-level understanding of the current water availability and quality issues and should promote proactive strategic actions instead of ignorance for the country to avoid a water crisis in the near future. Continued pressures from civil society and the scientific community can also play a role.

12.1.6 Running on Empty

The country's government has been aware of the country's national water deficit since 2004 and its water quality problems since 1979. South Africa is running out of water mainly due to continued over-exploitation of its water system and little attention being placed on the continued degradation of the already limited water resources. The Institute of Security Studies released a report in 2014 indicating that the country is overexploiting water resources on a national level and will continue to do so. Reconciliation strategies were also found to make little difference to the problem. Main driving forces are once again dilapidated wastewater treatment, high percentage of non-revenue water mostly through physical leaks as well as climate variability and continued water degradation across the country.

Recent warnings have all been dire water shortages which do not take into account the additional limitations of quality imposed on quantity. It has recently been estimated that South Africa will reach an annual water deficit of 17% or 38 billion m^3 by 2030. This is mainly due to huge water losses and poor infrastructure. The country's population also has an insatiable water appetite as it consumes 237 L per capita per day as opposed to the global average of 137 L per capita per day. The loss of ecological infrastructure such as wetlands also affects water availability and negatively affects the ability of water systems to offset flow in times of floods.

The new reality which will be faced by the country's population is higher cost for water and a significant cut in water use. The country will also have to invest in equitable allocation of water resources as well as effective infrastructure management, operation and maintenance. The current investment into sanitation infrastructure is still not enough to address the massive backlog of building new infrastructure and refurbishing the old. Focus also needs to be placed on moving away from flush lavatories to waterless sanitation, use of new water resources such as groundwater and reuse of wastewater. The country can therefore not settle for "business as usual" and needs to place more focus on the pollution of water resources as these will have additional impacts and may cause the water deficit to become more imminent.

12.2 Ultimate Losses and Costs

The continued degradation of water resources, intermittent supply and the overall increase of water stress will have cumulative effects on all sectors and the country's population as a whole. The poor and vulnerable of the country will however be mostly affected as they do not have the necessary resources to adapt or cope with the continued degradation of water resources.

The country's population, especially the poor, is affected by poor water quality and continued pollution by direct contact with contaminated water, indirect contact through contaminated food as well as inadequate healthcare facilities. The direct contact of polluted water is widespread across the country as clearly illustrated in

Chaps. 9–11. The present inadequate or absence of infrastructure in rural areas causes rural communities to make use of water which has been largely been contaminated by inadequately treated wastewater.

Most rural communities plant food close to polluted streams and fill their irrigation buckets with polluted water. Some communities especially those located close to mining activities or within close proximity of streams draining the Witwatersrand goldfields have to deal with water that has been contaminated by mine residue which includes toxic heavy metals and even uranium. Food grown in these affected communities may, therefore be contaminated by both microcystin and heavy metals associated with AMD. This risk is however unquantified. Consumers of food which buy their food from informal or formal economy may also be unaware that some produce may be contaminated by microcystin or sewage-borne pathogens. The potential of microcystin contamination and pathological loads are potentially high for food grown on commercial or subsistence farms near dams and rivers which have been polluted with sewage.

Healthcare facilities especially those located in rural areas are also unlikely to be equipped to diagnose and treat patients which may be suffering from central nervous system disorders in which environmental toxicity may have played a role. This is a new field of science in South Africa which is currently underresearched and poorly funded. The poor would and should, therefore, be the biggest beneficiary of improved water management as well as continued scientific research.

Increased pollution and costs of water pollution will also further restrict the access of the poorer sections of the population to safe drinking water and ultimately exacerbate risks associated with epidemic disease. Financial losses will, therefore, be incurred through absenteeism amongst employees, but also social consequences associated with loss of wages and loss of wage-earning family members as a consequence of fatal illnesses such as cholera. Once again, these impacts are felt the most by the poorest or most vulnerable sections of the country's population. The country already suffers from high incidence of morbidity from TB–HIV–Aids and the continued exposure to water which possibly contains drugs, hormones or toxicants can likely worsen the health of those already suffering from these conditions.

The country's tourism sector may also be affected negatively. Currently, South Africa benefits significantly from tourism and should water quality decline with associated changes in the safety of potable water, tourism may also decline. Recreational activities such as swimming and other water-related recreational activities could be seriously affected and tourism areas would consequently suffer a loss in revenue. Along with the loss of revenue, employment opportunities could also be negatively affected. A loss of employment cannot be afforded as the country already has an unemployment rate of 26.7%.

Additional economic ramifications should also be considered. The cost implications of poor water quality on industries, such as bottling and brewing industries, as well as on agricultural activities where algal blooms can clog irrigation systems as well as the decrease in quality of crops especially those of export value should be placed on the agenda. Overseas importers are unlikely to accept products where irrigation water has been compromised by wastewater effluents which may well be

12.2 Ultimate Losses and Costs

the case for the Berg River downstream of Paarl and Wellington as well as for the Vaal River downstream of the Barrage. Other examples may also exist across the country.

The use of contaminated water can also be accompanied by poor yield and/or contaminated crops which could increase food prices, while employment losses with decreased income place substantial pressure on the economy. Bottling and brewing companies may face increased costs, which would also be passed to the consumer and poorer sections of the population would be most highly affected.

Therefore, living with water quality issues such as eutrophication will have significant and serious consequences for the country's revenue and economy, with impacts ultimately transcending to many socio-economic sectors and negatively affect the society at large.

12.3 Possible Solutions and a Way Forward

South Africa's current usage trends will cause the country to have a water deficit of 17% by 2030 and this shortage will be worsened by climate change. All sectors and populations of the country will be at risk and it is therefore of utmost importance that the country as a whole is aware and understands the impact of actions on its already limited and scarce water resources.

Manners in which the country can possibly improve water quality and its overall water situation can be grouped into political action and financing, improved affordable technologies, policy reform as well as the incorporation of primary goals in collaboration with public, private and social sectors.

12.3.1 Political Will

The first step which needs to be taken which is also the most cost-effective and influential is to honour *Plakkaat 12* and prevent contamination before it occurs. This is however unrealistic as the country's infrastructure is currently in decay, poor public finances as well as the overall lack of political will. Sewage works should preferably in an ideal world be upgraded to ensure that facilities can cope with increased hydraulic loads. It is estimated that South Africa will need R30 billion which is simply not available. Tax revenues allocated to water management, therefore, have to increase and be used more efficiently. Due to the maintenance backlog, the DWS estimated that R80 billion a year over the next decade is needed to address water infrastructure decay.

The government is not currently fulfilling its responsibilities to provide enough clean and safe drinking water. South African National Standards water quality guidelines have also been diluted by government officials to mask the presence of certain metals, toxins and pathogens and are increasingly engaging in excessive chlorination in the hope to destroy pathogens not removed by treatment plants. Excessive chlorination consequently leads to further decay into a carcinogenic substance known as trihalomethane. This once again emphasis why transparent processes with accompanied high levels of shared information and coordinated actions are necessary. Responsible government stakeholders need to make full disclosure of all pathogens, microcystins and metals. This will, in turn, assist in the empowerment of the public, civil society organisations to hold elected officials accountable, address current risks and help with overall awareness and behavioural change.

12.3.2 Cost-Effective Technologies

Even though the country currently lacks the overall political will, low-cost methods to improve water quality exist. Hydro-Chemical Activation (HCA), a South African invention new technology, could be seen as a viable alternative to a complete and costly retrofit of the country's sewerage works. Costs are however site-specific but can be cheaper than expanding an existing plant because of improvements to efficiency. HCA is part of Advanced Pxidation Processes (AOPs) a set of chemical treatments that remove pathogens by exposing them to super-oxidants such as ozone and hydrogen peroxide, and sometimes ultraviolet light. The technology is however still too new for mainstream use but can find traction.

Activated carbon can also be considered as a viable solution to improve water quality. It is a carbon that has been altered with small pores that enlarges its surface area to adsorb toxins. This system can be easily retrofitted into South African potable water treatment plants to remove microcystin. After any uranium and other heavy metals entering a water treatment plant are precipitated and become part of the sludge, municipalities running these plants can use activated carbon in the final phase of the water purification process. Individuals can also buy various commercial products such as activated carbon cartridges that can be attached to taps. This technology is continually getting cheaper and with the implementation of innovative funding models, it can become affordable for the poor.

Hardware tools which could also be considered as investment opportunities could also include the following:

- **Urine Diversion Dehydration Toilets (UDDT)**: These are simple, low-cost amd on-site sanitation facilities which make use of dehydration processes possible for hygienically safe-on-site treatment of human excreta. These can be used in urban, semi-urban and in rural contexts, at household as well as institutional level.
- **Greywater Towers**: Cylindrical structures which are made of simple plastic bags or clothes, reinforced with wooden poles, filled with a soil mix and anchored into the soil. Vegetables can be grown in the holes cut in the sides of the bag. Greywater is then poured daily into the bag. These towers have been installed in the contexts of the Resource-Oriented Sanitation concepts in peri-urban areas in Africa on household level in Ethiopia and Uganda.

12.3 Possible Solutions and a Way Forward

- **Co-composting on a Large Scale**: Large-scale composting of organic waste has been used as a means to address immediate and serious environmental problems faced by governments of many African cities through the management of urban solid waste. In South Africa, the policy formulation and outline of work towards sustainable development has been done on a national level but direct implementation and formulation of the strategy lies with the provincial and local authorities. Composting is one method which can be used to recycle waste and reduce the amount of waste going to the landfill.

The reuse of WWTWs effluent should also become a necessary condition for any new WWTW under design. This will assist the country in terms of its limited water resources. For this to be possible in practice, the population's environmental awareness specifically in terms of the importance of protecting water resources as assets needs to be improved upon. Energy savings, as well as the quality of wastewater treatment equipment, can also be a driving force for advanced treatment.

12.3.3 Policy Improvement

South Africa is known for its excellent policy and legal frameworks. The implementation thereof is however in most cases weak due to political constraints as well as other factors.

The goal of policy reform should, therefore, be to encourage Private–Public Partnerships (PPPs) in the water sector. It is mainly needed in the re-invigoration of the national eutrophication programme. Sufficient funding as well as required technical skills to monitor and measure multiple points over long time periods is needed. Funding should also be expanded for the remote sensing of cyanobacterial blooms linked to credible public reporting. Satellite remote sensing can give accurate near real-time reporting capabilities over large geographic areas.

A fully funded multidisciplinary national research programme on the ß-methylamino-L-alanine (BMAA) microcystin is also needed as it is a risk for everyone, but especially the poor. A proper quality assurance programme should also be launched by the processing and distribution industry to ensure that they prevent contaminated food from entering the value while also educating consumers about possible risks.

12.3.4 Future Actions

The World Wide Fund for Nature South Africa (WWF-SA), supported by The Boston Consulting Group (BCG) established primary goals in collaboration with public, private and social sectors. These main goals include the following:

1. Increasing water consciousness by increasing sufficient knowledge and skills in the water sector;
2. Implementation of strong water governance through resilient stakeholder partnerships which advance the second phase of the National Development Plan to achieve water security under increased climate variability;
3. Managing water supply and demand regulations more strictly and protect water resources; and
4. Promoting and becoming a water-smart economy as well as a leader in Africa in commercialising low-water technologies especially for industry and agriculture (WWF 2017).

The group consequently identified six "no regret" actions which will have a high impact and be feasible to implement. These actions could significantly shape the future of water in South Africa and include the following:

- Improving social awareness on the seriousness of water scarcity in schools, business and communities through campaigns and social media platforms;
- Developing skilled employment opportunities, new enterprises as well as capabilities to effectively maintain green and grey water infrastructure and reduce losses;
- Innovative co-financing to maintain and protect ecological green infrastructure and combating further water losses from alien vegetation;
- Implementing a water pricing model to strategically differentiate tariffs with continuous growth in water demand, urbanisation as well as population growth;
- Commercialisation and implementation of at scale water reuse and improved irrigation efficiency technologies; and
- Increasing the access to information to share clearer understanding of water users' impact on water and so doing advancing collective action (WWF 2017).

Importantly, collective will play a vital role in building a sustainable water future for all stakeholders. The country needs to collaborate to mitigate risks, create and seize opportunities as well as place the preservation and maintenance of its water resources at the top of its agenda for it to create a water-secure future. A massive drive is also needed to improve performance as well as accountability, especially in public sector water institutions and national and local government. Water stewardship, as well as collective action with the required understanding of water supply and demand, is therefore needed for the country to try and shape an improved and sustainable water future.

12.4 Conclusions

Water quality can, therefore, be considered to be a good measurable indicator of government management within South Africa. Current trends indicate that the government has lost faith in its ability to implement and deliver under more challenging conditions such as increased effects of climate variability. The increased involvement

12.4 Conclusions

of the private sector has consequently been encouraged and has been considered to be the key to stronger governance for the water system.

Some steps have already been taken towards increasing greater water consciousness especially during the 2016–2017 droughts. High-level understanding is the key in trying to understand how partnerships can build necessary skills and competence which the country's water sector desperately needs. The country needs to develop an overall competence which emphasises that more can be done with less in an effort to reduce the already above average water use especially in terms of domestic use. Industries, as well as agricultural sectors, also need to invest in more water efficient technologies to reduce their water demand and assist in decreasing the country's water deficit.

All of the mentioned challenges can be converted into opportunities. Real opportunities exist where South Africa can emerge as a leader in the transition into water-smart economies. This can be achieved with the investment into new cost-effective technologies as well as enterprise innovations which all aim to contribute to ensured water security. Emphasis should also be placed on having quality wastewater treatment which can be reused and recycled to be used for the irrigation of crops to ultimately try and ensure future sustainability. Decisive steps, however, need to be taken now as the country cannot afford to wait until a next national disaster such as the current drought.

References

Bugan RDH, Jovanovic NZ, de Clercq WP (2015) Quantifying the catchment salt balance: an important component of salinity assessments. S Afr J Sci 111:1–8

Camden-Smith B, Pretorius P, Camden-Smith P, Tutu H (2015) Chemical transformations of metals leaching from gold tailings. Paper presented at the 10th international conference on acid rock drainage and International Mine Water Association annual conference, Chile

Casas I, de Pablo J, Gimenez J, Torrero ME, Bruno J, Cera E, Finch RJ, Ewing RC (1998) The Role of pe, pH and carbonate on the solubility of UO_2 and uraninite under nominally reducing conditions. Geochim Cosmochim Acta 62:2223–2231

Cillie GG, Odendaal PE (1979) Water pollution research in South Africa. J WPCF 51:458–465

Coetzee H, Winde F, Wade PW (2006) An assessment of sources, pathways, mechanisms and risks of current and potential future pollution of water and sediments in gold-mining areas of the Wonderfonteinspruit catchment. WRC Report, No. 1214/1/06. Water Research Commission, Pretoria

Conradie KR, Barnard S (2012) The dynamics of toxic microcystis strains and microcystin production in two hypertrofic [sic] South African reservoirs. Harmful Algae 20:1–10

Davidson IO (2011) Written response by the Minister of Water Affairs to Internal Question Paper 04 (7 Mar 2011). South African National Assembly Question 574. NW477E

De Villiers S, Thiart C (2007) The nutrient status of South African rivers: concentrations, trends and fluxes from the 1970s to 2005. S Afr J Sci 103:343–349

DEA (Department of Environmental Affairs) (2012) South Africa environment outlook. Chapter 4: Inland Water (Draft 2) 18 Jan 2012. DEA, Pretoria

Downing TG (2011) The development of an analytical system for ß-N-methyylamino-L-alanine and investigation of distribution of producing organisms and extent of freshwater contamination. Water Research Commission Report 1719/1/10. WRC, Pretoria

Downing TG, van Ginkel CE (2004) Cyanobacterial monitoring 1990–2000: evaluation of SA data. Water Research Commission Report 1288/1/04. WRC, Pretoria
DWA (Department of Water Affairs) (2013) National eutrophication monitoring programme phosphorus and chlorophyll summaries. DWA, Pretoria. Available from: http://www.dwaf.gov.za/iwqs/eutrophication/NEMP/report/NEMPyears.htm
Economics Plus (2010) The South African water crisis: an economic impact study. Plus Economics, Pretoria
Harding WR, Paxton B (2001) Cyanobacteria in South Africa: a review. Water Research Commission Report TT153/01. WRC, Pretoria
Kalyan SV (2010) Parliamentary Query 2082 to the Minister of Water and Environment Affairs (Internal Question Paper 19). South African National Assembly
Matthews MW, Bernard S (2015) Eutrophication and cyanobacteria in South Africa's standing water bodies: a view from space. S Afr J Sci 111:1–8
Oberholster PJ, Ashton PJ (2008) State of the nation report: an overview of the current status of water quality and eutrophication in South African rivers and reservoirs. CSIR, Pretoria, pp 1–14
Oberholster PJ, Botha A-M, Myburgh JG (2009) Linking climate change and progressive eutrophication to incidents of clustered animal mortalities in different geographical regions of South Africa. J Biotechnol 8:5825–5832
Scott WE (1991) Occurrence and significance of toxic cyanobacteria in southern Africa. Water Sci Technol 23:175–180
Turton AR (2013) Debunking persistent myths about AMD in the quest for a sustainable solution. Paradigm Shifter 1. South African water, energy and food forum, Johannesburg
Van Ginkel CE (2004) A national survey of the incidence of cyanobacterial blooms and toxin production in major impoundments. Report No.: N/0000/00/DEQ/0503. Department of Water Affairs and Forestry, Pretoria
Van Ginkel CE, Silberbauer MJ, Vermaak E (2000) The seasonal and spatial distribution of cyanobacteria in South African surface waters. Verh Int Ver Theor Angew Limnol 27:871–878
Winde F (2010a) Uranium pollution of the Wonderfontein Spruit, 1997–2008. Part 1: uranium toxicity, regional background and mining-related sources of uranium pollution. Water SA 36:239–256
Winde F (2010b) Uranium pollution of the Wonderfontein Spruit, 1997–2008. Part 2: uranium in water—concentrations, loads and associated risks. Water SA 36:257–278
WWF-SA (World Wildlife Fund South Africa) (2017) Scenarios for the future of water in South Africa. WWF-SA

Appendix
DWS Water Quality Guidelines

Physical water quality parameters

| | pH | | | |
	Water quality guidelines and standards (pH units)			
Water use	Ideal	Acceptable	Tolerable	Unacceptable
Domestic use	6–9	9–11	4–6	<4 and >11
Aquatic ecosystems	No guidelines			
Recreational	6.5–8.5	5.0–6.5 and 8.5–9.0	No guideline	0–5.0 and >9.0
Irrigation	6.5–8.4	No guideline	<6.5 and >8.4	No guideline
Industrial	7.0–8.0	No guideline	6.0–7.0 and 8.0–9.5	>9.5

| | Electrical conductivity (EC) | | | |
	Water quality guidelines and standards (mS/m)			
Water use	Ideal	Acceptable	Tolerable	Unacceptable
Domestic use	0–70	70–150	150–450	>450
Aquatic ecosystems	No guidelines			
Recreational	No guidelines			
Irrigation	40	40–90	90–540	>540
Industrial	0–15	15–30	30–70	>70

Chemical water quality parameters

Calcium (Ca^{2+})
Water quality guidelines and standards (mg/l)

Water use	Ideal	Acceptable	Tolerable	Unacceptable
Domestic use	0–32	32–80	>80	No guideline
Aquatic ecosystems	No guidelines			
Recreational	No guidelines			
Irrigation	No guidelines			
Industrial	No guidelines			

Chloride (Cl^-)
Water quality guidelines and standards (mg/l)

Water use	Ideal	Acceptable	Tolerable	Unacceptable
Domestic use	0–100	100–200	200–1 200	>1 200
Aquatic ecosystems	0.0002 mg/l (0.2 µg/l)	No guideline	0.0035 mg/l (0.35 µg/l)	0.005 mg/l (5 µg/l)
Recreational	No guidelines			
Irrigation	100	<140	140–700	>700
Industrial	0–20	20–50	50–120	>120

Sodium (Na^+)
Water quality guidelines and standards (mg/l)

Water use	Ideal	Acceptable	Tolerable	Unacceptable
Domestic use	0–100	100–200	200–600	>600
Aquatic ecosystems	No guidelines			
Recreational	No guidelines			
Irrigation	70	70–115	115–460	>460
Industrial	No guidelines			

Nitrate (NO_3^-)
Water quality guidelines and standards (mg/l)

Water use	Ideal	Acceptable	Tolerable	Unacceptable
Domestic use	0–6	6–10	10–20	>20
Aquatic ecosystems	<0.5	0.5–2.5	2.5–10	>10
Recreational	No guidelines			
Irrigation	5	No guidelines	5–30	>30
Industrial	No guidelines			

Sulphate (SO_4^{-2})
Water quality guidelines and standards (mg/l)

Water use	Ideal	Acceptable	Tolerable	Unacceptable
Domestic use	0–200	200–400	400–600	>600
Aquatic ecosystems	No guidelines			
Recreational	No guidelines			
Irrigation	No guidelines			
Industrial	0–30	30–80	80–150	>150

Ammonia (NH_4^+)
Water quality guidelines and standards (mg/l)

Water use	Ideal	Acceptable	Tolerable	Unacceptable
Domestic use	0–1	1–2	2–10	>10
Aquatic ecosystems	<0.5	0.5–2.5	2.5–10	>10

(continued)

(continued)

Ammonia (NH_4^+)
Water quality guidelines and standards (mg/l)

Water use	Ideal	Acceptable	Tolerable	Unacceptable
Recreational	No guidelines			
Irrigation	5	No guidelines	5–30	>30
Industrial	No guidelines			

Phosphate (PO_4^{-2})
Water quality guidelines and standards (mg/l)

Water use	Ideal	Acceptable	Tolerable	Unacceptable
Domestic use	No guidelines			
Aquatic ecosystems	<0.5	No guidelines		
Recreational	No guidelines			
Irrigation	No guidelines			
Industrial	No guidelines			

Biological/microbial water quality parameters

Chlorophyll a
Water quality guidelines and standards (µg/l)

Water use	Ideal	Acceptable	Tolerable	Unacceptable
Domestic use	0–1	No guideline	1–15	>10
Aquatic ecosystems	<0.5	0.5–2.5	2.5–10	>10
Recreational	0–15	No guideline	15–30	>30
Irrigation	No guidelines			
Industrial	No guidelines			

Faecal coliform
Water quality guidelines and standards (Counts/100ml)

Water use	Ideal	Acceptable	Tolerable	Unacceptable
Domestic use	0	0–10	10–20	>20
Aquatic ecosystems	No guidelines			
Recreational	0–130	No guideline	600–2,000	>2,000
Irrigation	1	No guideline	1–1,000	>1,000
Industrial	No guidelines			

PGMO 03/06/2019